CABLEVIEWING

COMMUNICATION AND INFORMATION SCIENCE

Edited by
BRENDA DERVIN
The Ohio State University

Recent Titles

CABLEVIEWING

Carrie Heeter and Bradley S. Greenberg

ABLEX PUBLISHING COMPANY
NORWOOD, NEW JERSEY

Printed in the United States of America

Library of Congress Cataloging-in-Publication Data

Heeter, Carrie.
  Cableviewing / Carrie Heeter and Bradley S. Greenberg.
    p. cm.—(Communication and information science)
  Bibliography: p.
  Includes index.
  ISBN 0-89391-466-5. ISBN 0-89391-467-3 (pbk.)
  1. Television audiences—United States. 2. Television views—
United States. 3. Cable television—United States. I. Greenberg,
Bradley S. II. Title. III. Title: Cable viewing. IV. Series
HE8700.66.U6H44 1987
384.55′56′0973—dc19   88-10450
                                                                CIP

Ablex Publishing Corporation
355 Chestnut Street
Norwood, New Jersey 07648

Contents

        Bradley S. Greenberg, Roger Srigley, Thomas F. Baldwin, and
        Carrie Heeter

Chapter 21   Conclusions and a Research Agenda   **289**
        Bradley S. Greenberg and Carrie Heeter

        References   **306**

        Author Index   **313**

        Subject Index   **316**

Acknowledgments

Hundreds of individuals have been part of this volume—as researchers, interviewers, coders, data analysts, and literature searchers. Literally thousands have been interviewed, as the individual sections will show. Without both groups, there would be nothing to report here. So we have many to whom we are grateful.

Financial support for different studies or for the support of different persons associated with the studies has been obtained from Coaxial Communications of Columbus, Ohio, the College of Communication Arts and Sciences at Michigan State University, Continental Cablevision, the Communication Technology Laboratory, Horizon Cable, the Departments of Communication and Telecommunication at MSU, the National Association of Broadcasters, the National Science Foundation, the Michigan State University Foundation, Rainbow Corporation, and United Cable.

Special cooperation was obtained from the Arbitron Corporation and the Plymouth, Michigan, school system.

Preparation of the manuscript was significantly facilitated by the excellent staff work of Ann Alchin, Cynthia Brown, Cindy Shinaberry, Ann Spalding, Brenda Sprite, Bonnie Sturdivant, Debbie Tigner, and Ann Wooten.

This work is dedicated to the many students who have worked in the Communication Technology Laboratory, and to the Greenberg women—Dee, Beth, Shawn, and Debra, without whom it would not have been worth the pains.

About the Authors

(All authors were affiliated with Michigan State University at the time these studies were conducted.)

Kwadwo Anokwa is a Ph.D. candidate in the Mass Media Program at Michigan State University.

David Atkin is a faculty member at Southern Illinois University, Carbondale, IL.

Thomas F. Baldwin is Professor of Telecommunication and Chair of the Mass Media Ph.D. Program at Michigan State University.

Ed Cohen is Manager of Audience Measurement and Policy Research with the National Association of Broadcasters in Washington, DC.

David D'Alessio is a faculty member at Grand Valley State College, Grand Rapids, MI.

Bradley S. Greenberg is a Professor of Telecommunication and Communication at Michigan State University.

Carrie Heeter is Director of the Communication Technology Laboratory at Michigan State University.

Carolyn A. Lin is Assistant Professor of Radio and Television at Southern Illinois University.

Carol Mackey is an Account Executive with AT&T in Southfield, MI.

Julie McDonough is a stained-glass artist.

D. Stevens McVoy is Vice-President of Coaxial Scientific in Columbus, OH, and Adjunct Professor of Telecommunication at Michigan State University.

Ronald Paugh is Assistant Professor and Director of Research at the Gill Center, Ashland College, Ashland, OH.

Sherri Sipes is a Media Supervisor with the Leo Burnett Advertising Agency in Chicago.

Roger Srigley is a Ph.D. candidate in the Mass Media Program at Michigan State University.

Cynthia Stanley is an account executive for Sportsvision in Chicago, IL.

Foreword

David S. Bender

Vice-President for Marketing and Research
USA Cable Network

The "modern history" of audience measurement of advertising-supported cable networks began with a cable network that isn't strictly a cable network. In 1976, Ted Turner began to uplink the signal from his WTBS, Channel 17 in Atlanta, to RCA Satcom I, which satellite in turn made the station's signal available to cable operators all over the country. Turner launched this "superstation" to make money, and the source of revenue was to be national advertising. By the late 1970s, of course, advertisers and their agencies had had a great deal of experience buying television time, and the process had become a "science" inasmuch as it had been quantified. Television time had become too expensive for any single advertiser to sponsor an entire primetime program on one of the three broadcast networks; 60- and 30-second commercial units were the rule. In very large measure, the value of these spots was determined by the total number of households and persons tuned to a program. But "All in the Family" commanded more money for each one thousand viewing households than lesser-rated programs because it delivered more total households. It is difficult to know the reasons one commercial in "All in the Family" seemed to advertisers and agencies to be worth more than airing two in a program seen by half as large an audience, but there are several possibilities:

1. The age/sex demographics and/or the characteristics of viewing households for "All in the Family" may have more closely matched an advertiser's target audience than did other programs. The notion here is that all households and persons are not equally valuable to a particular advertiser.

2. While two spots in a program rated half as high as "All in the Family" indeed would deliver as many household impressions, one spot in "All in the Family" would doubtless deliver more *different* households. That is, one household can't see one commercial twice; but if the same commercial runs twice in a program, one household *can* see it twice. Advertisers often have as a goal

reaching as many different households as possible, and higher-rated programs deliver greater "reach."

    3. "All in the Family" was an enormously successful program, and advertisers may have wanted to associate with its popularity and success. While not quantifiable, this factor carries with it an added value.

    4. It's easier for an advertising agency to buy one spot than two. Or fifty than one hundred.

    5. "All in the Family" was truly a unique program—one of a kind—and this low supply no doubt created high demand, and, in turn, high prices.

The Super Bowl has become perhaps the best example of a program which commands more money per thousand viewers than other programs. It meets all five of the above criteria in spades.

WTBS entered a national television advertising sales market which had been structured to devalue, if not exclude, entities like WTBS every bit as much as the NFL is structured to exclude good high school quarterbacks. How to succeed in that environment? WTBS took the position that, when in Rome. . . .

From the outset, WTBS and, since then, other cable networks, wanted to have viewership measured *exactly as the viewership of broadcast networks is measured*. This goal seemed to cable networks to be a way—the only way—to assure advertisers and agencies that audiences to cable were *certifiably* of some size. (It is important also to know that the cable networks have always gone beyond mere measurements of gross household audiences. Further measurements are and have been our stock in trade; but the base measure has been the number of households delivered for an advertiser.)

To play on the same field as the three broadcast networks meant viewing to cable networks must be measured using Nielsen's audimeter sample. In the late 1970s and early '80s this nationwide panel comprised about 1,250 households. In February 1981, 19% of these metered households were able to receive WTBS. Those 238 households provided national household ratings for the "superstation." And the goal had been achieved: Household viewership of WTBS was being measured by the same yardstick used to measure ABC, CBS, and NBC.

The relatively low ratings produced for WTBS, along with its number of potential viewing households, made literally *all* broadcast network programs more valuable than WTBS's best-rated shows.

As I write, in early 1987, 17 cable networks have contracted with Nielsen to measure their viewership using the meter sample. All but two (HBO and Showtime, which don't sell advertising) use the data to certify to advertisers the size of their household audiences nationwide.

Two other noteworthy, early developments:

- In 1983 Nielsen formed a new division, Nielsen HomeVideo Index, and charged it with providing measurements of cable viewership and of VCR use.
- In 1983 Nielsen increased the size of its audimeter sample from 1,200 to

1,700, largely to improve its ability to measure viewing to cable networks. This immediately made several networks eligible—the criterion being a minimum number of metered households able to receive them—for measurement by meters.

Here I must pause to explain that, while the majority of advertising dollars spent in network cable was tied to guarantees to be verified by Nielsen meters, other, quite different sales techniques were sometimes used. The smaller (in terms of distribution) cable networks, neither widely viewed nor even eligible for meter measurement due to low penetration of the sample, relied for some years on what became known as the "concept sell." MTV, before Nielsen began to measure it in late 1983, had very good advertising sales success based on its sales staff being able to ask agency buyers the rhetorical question, "Who do you *think* is watching this stuff?" Little old ladies? Of course not: It's viewed by persons aged 12–34, exclusively. (It didn't hurt that most media buyers were members of that demographic group, watched MTV, and were certain their parents never would.) The MTV approach worked quite well *without* the support of continuous viewership measurement reported regularly. Other networks used the same tactic: Arts (which became Arts & Entertainment in early 1984) was able to sell advertisers upscale, well-educated, cosmopolitan viewers. Who else after all, would watch ballets, operas, and symphonies? And everyone knows the best way to attract men is to program sports, so ESPN was able to attract advertisers before it was measured regularly by Nielsen.

In the early 1980s several voices could be heard urging—warning—cable networks to beware playing the Nielsen ratings game. The voices came from the agency and advertiser side as well as from the cable network side. They had a good point: Cable network ratings are lower; the cable nets would always be at a disadvantage. But it was already too late. WTBS claimed the lion's share of cable network ad sales revenues in the early '80s, and the rest of us couldn't wait to have our own regularly issued metered ratings reports.

And so we got them. The count of cable networks by the years they first signed Nielsen meter contracts:

1980	1
1982	4
1983	3
1984	5
1985	0
1986	4

Cable networks, in 1986, equalled about a broadcast network and a half in revenues for Nielsen.

It would now be very difficult for a cable network such as USA Network to approach an advertiser or an agency without ratings, or to forge a deal without guaranteeing the delivery of a specified audience, to be verified by Nielsen ratings.

One mixed blessing of having Nielsen household ratings "just like the big-boy networks" has been that most agencies now put together media plans which include cable as well as broadcast networks in nearly every national television campaign; the mixed part of the blessing is that we're included in these plans because our audiences can now be quantified, but the instant the quantifiability went away we'd find our networks excluded from the media plans. And a medium which has not been included in a plan is rarely bought.

So as long as we continue to feed ratings into the agency slot machines, we've got a fair chance of being included as a part of national buys; if the ratings flow were to stop, however, we'd be allowed only to stand and watch as someone else (probably in barter syndication) pulled the handle.

The early goal was to be measured just as other national television is measured, to play on a level field. That has been accomplished, but only for household ratings. The Nielsen audimeter measures only two things: whether the set is on or off, and which channel it's tuned to when it's on. But sets don't watch TV, nor do households. People do. And advertisers know that households don't buy automobiles, deodorant, or beer; people do. Networks—cable and broadcast—sell people to advertisers. (Perhaps "rent" people would sound less callous?) Marketers of products and services are interested in communicating with people targeted as their best potential customers. A television ratings research company which does not measure viewership by individual persons hasn't a prayer; very little television is bought based on households; almost all of it is bought to reach a specific demographic group.

The measurement of persons viewing has become the most controversial issue in television research in the 1980s. Cable, of course, is spank in the middle of the fray.

For a generation, Nielsen has measured persons viewing broadcast network programs by using its National Audience Composition (NAC) sample. This panel sample comprises about 2,700 households, about one-third of which record in diaries the viewership by persons of all sets in the household, quarter-hour by quarter-hour, a week at a time. About three-quarters of the weeks in each year are measured this way. Demo data are conformed to meter data; that is, each demographic group's concentration in the total audience of households is rendered as a ratio (of viewers per viewing household), then multiplied by the total number of households according to the meters.

Since only about 900 households participate at any one time, and since only about one-half of them have cable, and since cable household ratings are typically quite low—it is rare for 5% of the households which can receive a cable network to be watching it at the same time—there are simply not enough persons in the NAC panel to provide reliable demographic viewership data for cable networks. In fact, demographic data for broadcast networks provided by the NAC sample is unreliable enough that agencies have been clamoring for years to have the sample size increased.

How, then, to provide demographic data for cable networks? Nielsen found a way: There is a division of Nielsen called Nielsen Station Index (NSI) which measures viewership of over-the-air TV stations in individual markets around the country. Four times each year—in the February, May, July, and November "sweep" months—NSI mails 50,000 diaries weekly, and receives more than 25,000 back with a week's worth of viewership recorded. The 100,000+ completed diaries for each sweep month are sufficient to provide household and demographic data for each of the nation's 209 television markets. The 50,000+ diaries completed each sweep month by cable households are sufficient quite reliably to project demographic viewership of cable networks.

Since diary-keepers are less complete and accurate in recording their viewership of cable channels than of broadcast channels, the diary method understates viewing to cable, but since the NTI household meter records cable viewing every bit as well as it records viewing to over-the-air channels, the NSI diary-generated demographic data are "conformed" to NTI meter-generated household data, which raises the persons viewing projections to levels consistent with household projections.

This method of estimating demographic viewership of cable networks is not perfect (not that the NAC panel method is). It provides only 16 weeks worth of persons data yearly, while NAC provides 39.

Worse, persons' viewership of cable is being measured differently from that of broadcast networks (and from some syndicated programs which use the NAC panel, too), and not to be measured as the broadcast networks are measured has been a liability for cable networks.

Now I've described all that to lead up to writing that it's all going to change, and soon. By the time this book is published the entire system of national television audience measurement in the United States will have changed. The sweeping changes are due to the identification of a problem: current diary-based methods of collecting data about the demographic composition of television audiences are inadequate; and to the solution to that problem: the peoplemeter.

Aspects of Nielsen's diary system have been judged wanting. First, response error is high: Viewing to cable channels and independent stations is often incorrectly recorded or not recorded, while viewing to popular network programs is sometimes recorded although it did not occur. Diaries are typically completed for all members of the sample household by only one member (often the female head) of the household; and diary completion is many times undertaken post hoc. Second, the NAC diary sample (2,700 households, of which only one-third complete diaries in any single week) is simply too small to provide the reliability needed by buyers and sellers of advertising. The unreliability of the NAC demographic data can be illustrated by two examples: First, how do you suppose advertisers felt when Nielsen reported there were 583 women aged 18–49 for every 1,000 households viewing ''Santa Barbara'' during the first 2 weeks of April 1987, but only 418 women 18–49 per 1,000 households viewing the same program during the following 2 weeks? How do you suppose NBC felt? The network and its advertisers had lost 28% of their women 18–49 audience due to measurement of different thirds of the NAC sample. Similarly, during the same ratings periods, ''Pee Wee's Playhouse'' on CBS dropped from 1,238 to 939 children 2–11 per 1,000 viewing households. That's a 24% ''loss'' of target audience. Did these losses actually occur? Almost certainly not; that they occurred in the Nielsen sample suggests serious unreliability. Those are examples of the observation that a man with a watch always knows what time it is, but a man with two or more watches never knows what time it is.

Enter the peoplemeter, a one-watch television measurement system.

The peoplemeter sample promised by Nielsen is to grow from 2,000 households in September 1987 to more than 4,000 households in September 1988. AGB, Nielsen's immediate competitor, has promised 2,000- and 5,000-household samples by those respective dates. These samples will be several times the size of the present NAC samples, and that will ameliorate reliability problems.

The peoplemeter itself is a complex instrument which performs quite simple functions:

(1) like present set meters, the peoplemeter records whether each set in the sample household is on or off.
(2) like present set meters, the peoplemeter records which channel is tuned on each switched-on set.
(3) in addition, members of peoplemeter sample households are asked to record their personal viewership to any set in the household by keying their individually-assigned number either on a hand-held remote device or on a small box atop the television. This personal viewing information is stored along with set-tuning data, and is sent early every morning via telephone lines to Nielsen's computer.

Unlike paper-and-pencil diaries, the peoplemeter requires viewing be recorded coincidentally with the viewership itself; post hoc recording of viewership is eliminated.

Minute-by-minute (even second-by-second) data on personal viewing—contrasted with quarter-hour data collected in diaries—are available, making the measurement of viewership of commercials technologically possible.

Note that the current, diary-based system uses two separate samples: the NTI setmeter sample, which provides household data; and the NAC diary sample, which provides persons data. The latter data are conformed to the former data. Peoplemeter methodologies use only one sample, obviating the need to conform data from the diary sample to the meter sample.

In the abstract, cable networks should welcome the peoplemeter technology: It will provide continuous (all day, every day) measurement of both household and persons viewership. Data—especially persons data—will be available more quickly, and online, we've been promised. One large sample should entail less error than two smaller samples one of which is conformed to the other. Importantly (but also only theoretically), a person watching USA Network or another cable network has no more or less reason to have his or her button pushed—or not pushed—than a person watching ABC or another over-the-air station. The disadvantage to cable networks inherent with the diary methodology should disappear. Registration of personal viewership concurrent with the viewership should eliminate the overstatement of viewing to broadcast signals (due to familiarity with broadcast networks stations, and programs) and eliminate the understatement of viewing to cable channels (due to unfamiliarity and confusion over which cable channel is being watched).

All those facets of the peoplemeter promise to work out well for cable networks, but, to date (early 1987), cable networks have seen little peoplemeter data, and that from a sample of only 1,000 households.

Nielsen will abandon the 1,700-household NTI meter sample, beginning September 1987, in favor of a 2,000-household peoplemeter sample. Broadcast networks and widely syndicated programs will use this sample as the source of both household and demographic (persons) ratings. Cable networks will use the new sample as the source of only household ratings (NSI diaries will be conformed to these peoplemeter-based ratings for persons viewership estimates) from September 1987 through September 1988. As of the latter date, it is expected that, with a sample which will have grown to more than 4,000 households, Nielsen will be able reliably to use the peoplemeter sample to provide both household and persons data for cable networks (at least the large ones).

At that time, for the first time, cable networks viewership will be measured just as broadcast network viewership will be. Perhaps it would be reaching to say ''an era will have begun,'' or something else grand, but it can truly be said a goal will have been reached!

Arlo Guthrie once told a very long story to introduce another pretty long story, and I'm afraid I'm going to have to confess that's what I've done here. I've discussed Nielsen audience research, which is used by cable networks to support their arguments that they're good advertising vehicles. Nielsen (or its competitors AGB and ScanAmerica) aren't the only research companies used by cable networks, and meters and diaries aren't the only research tools we use. But as I look at my previous sentence, I feel my nose growing a little, because that kind of syndicated research is just about all we do. The reason for that circumstance is that Nielsen uses up the lion's share of our research budgets. I argue every year—and I know my colleagues at other cable networks do, too—for an allocation of research dollars which would be spent on programming, marketing, and affiliate system studies. But I pretty much lose every year. The economics of my company—and again, I believe, of other cable networks—don't permit us to do much primary (that is, not syndicated; initiated by us and proprietary to us) research.

Thus it is with glee, delight, and even a bit of a larcenous feeling that I welcome this volume. I happen to know Professor Greenberg enjoys wine, and especially wine from California, and most especially the wines produced by small vineyards there. The economics and techniques of "boutique" wineries are quite different in scale and in kind from the economics and techniques at Paul Masson or Christian Brothers. And what a blessing that is for lovers of good, innovative wine!

I suspect Professor Heeter also enjoys good wine produced to exacting specifications on budgets dwarfed by Taylor or Inglenook. (I suspect this only because, if she doesn't, the analogy won't be as meaningful.) Professors Greenberg and Heeter have produced a corpus of "boutique" research for this volume which brings to mind the best California wines I've sampled.

- Not too many people have worked on this research, and that's a plus; all the researchers know about all the research. That's hardly the case at Nielsen, whose clients have often complained the right hand has apparently never even ridden in the same elevator bank as the left hand.
- The research in this volume wasn't all done at the same time, didn't use the same sample or couple of samples, and wasn't conducted to please clients or to raise corporate profits above last year's.
- Heeter and Greenberg's research program didn't measure the same things again and again. I used to work for Louis Harris, who often advised our clients that, while observing trends in research indeed has great value, when a trend is found not to measure the appropriate phenomenon or is found to measure it inadequately, it is time to start a new trend line. Anyway, there are opportunities to measure viewership variables much more interesting than those regularly tracked by Nielsen, and this is the key to the value of the research presented here.
- These topics would be fascinating to students of television viewership even if they didn't focus on cable; even if they weren't well designed and executed. (They do, and they are.) Nobody in the commercial television industry has done much research on viewing styles, but they should. One or more of the broadcast networks may have conducted a study or two, but any such studies haven't been released; and they can't have helped the nets much anyway, considering the precipitous, continuing decline of network ratings and shares in all dayparts.
- To have this series of cable-specific studies is valuable to me, and it will have value for others who work in cable research, as well. We haven't had many clues about the process of program selection. All we know is that some percentage of households and persons who could have watched USA Network or

another channel did or didn't do so. Why? What was the second choice? How are pay cable households different from basic cable-only and noncable households in the processes used to choose programs?

- Heeter and Greenberg have assembled data based largely on people, not households. Nielsen's samples (and AGB's and ScanAmerica's) are *household* samples; households, not individuals, are selected. That's true of the new peoplemeter samples, too. Persons are included in the sample because they live in sampled households. But it's the people, not their homes, which are important to buyers and sellers of advertising. An important feature of this volume of research is its concentration on individual, rather than collective, behavior.

Earlier, I mentioned cable networks had been warned not to try to play the Nielsen ratings game against the broadcast networks. Economic exigencies have forced all of us to do exactly that, and we've had good success. I can't envision a day when we won't have to certify our delivery of audiences for advertisers and agencies.

But I can see a time when researchers at cable networks (and cable systems and MSOs) will be able to do more programming and marketing research than we do now to benefit our companies and our viewers. I am able to envision that time—however far in the future it might be—because this volume provides a tantalizing look at what it will be like. Sort of like a Gallo drinker getting a taste of what the 1992 Ridge zinfandels will be like.

David C. Bender
New York

## CHAPTER 1

Introduction: A Context for Studying Cableviewers

Carrie Heeter and Bradley S. Greenberg

Given our ever-changing media environment, this book will soon become an historical documentation of how people watched television during the decade when cable television broke ABC, CBS, and NBC's total domination of TV viewing. Even though there will be ongoing change as new technologies become embedded in our media habits, many of the changes wrought by cable will persist and influence future media generations. For example:

- Cable, with its multitude of channels, has broken the network oligopoly.
- TV is no longer an economic "public good" that anyone with a TV set can watch. Instead, subscribers pay to receive channels.
- Some channels now specialize in a particular type of program.
- People in the same city (and across cities) no longer have access to the same set of programs.
- Cable subscribers more often also have remote control selectors.

Cable television entered the picture as early as the 1950s, but only recently have the right technological, regulatory, and market factors combined to allow cable to dramatically alter the structure of television in the U.S. It began as Community Antenna Television (CATV), using tall antennae to pick up and distribute rural broadcast signals too weak to be received from household antennae, often allowing reception of one or a few channels in locales where no channels had been available.

In 1963, the FCC decided it was in the public interest to protect existing broadcasting facilities. To prevent cable from "destroying" local stations and rural coverage by importing distant broadcast signals, cable systems using microwave relays were required to carry local stations and to avoid any duplication of programs on local stations. FCC limitations on cable continued in 1966, when the Supreme Court ruled that cable was "reasonably ancillary to broadcasting," thus placing it under FCC jurisdiction. Up to this time, cable was exclusively a distributor of broadcast channels. The FCC maintained that limited role for cable by enacting regulations to prevent cable systems from originating any programming.

In 1969, the FCC changed its posture toward cable, recognizing the potential of the medium to increase the diversity of programming available to viewers. Thirty-six or more channels, rather than the 12 or fewer of previous cable systems, had become technologically possible. Cable companies were now required to originate programming. More rules were added in 1970. Although these rules were eventually challenged and overturned in 1977 as violations of the First Amendment, the regulations served to inhibit the growth of pay cable until they were rescinded. Advertising was prohibited, and the proportion of movies, sports, and series that could be carried on pay cable was restricted.

In 1972, the FCC introduced regulations which mandated that any new cable system provide public access channels, local origination, and a minimum channel capacity. Deregulation has occurred progressively and gradually since 1972. In 1977, the FCC regulations against pay cable were found to be in violation of the First Amendment. That same year, the FCC also authorized use of 4- to 5-meter earth stations (smaller and less expensive than those previously allowed) to receive satellite signals. The combined impact was facilitation of the development of satellite distribution of nonbroadcast channels, both pay cable and advertiser-supported channels. Growth in that area was so dramatic that, within 5 years, the principal function of cable systems was no longer retransmission of broadcast signals. Continued deregulation of cable has included dropping the "must carry" rules which required carriage of local broadcast stations on a cable system.

Ironically, although cable television began by providing broadcast reception for rural areas, reducing the rural–urban gap in channel availability, it now serves to magnify those inequities. In 1981, more than 70% of all cable systems carried only 12 channels. Newer systems being built in urban areas all had more channels, and profitable smaller systems were gradually upgraded. Rural and small towns were among the first to be wired for cable, leaving them with 12 channels rather than 26- or 54-channel systems. Together with the geographic differences, there are vast differences in availability of programming between cities and within individual cities, since cable systems are programmed and operated by local franchises, each choosing its own channel line-up.

Remote-control penetration, while not linked only to cable, occurs more often in cable households, since remote control units are frequently made available by the cable system. During March 1986, 56% of basic cable households and 31% of noncable households were equipped with remote control devices (Thompson, 1986). By the year 2000, an estimated 90% of households will have remote control (Market Opinion Research, 1987).

Measuring cable viewership is much more complex than measuring broadcast viewership used to be. Traditional diary methods, which ask subjects to identify which channels they watch and at what time each day, become untenable when there are more channels than viewers can remember by name. Identifying channels to assess viewership gets very confusing when, for example, broadcast channel 6 (the local CBS affiliate) is assigned to cable channel 2, while cable channel 10 is a distant CBS affiliate. Viewership has traditionally been measured in quarter hours. With cable and with remote control, quarter hours are no longer an adequate interval to account for the variety of channel changing which might occur in a 15-minute block. Metered Nielsen households allow continuous measurement of household viewing across a national sample, but a major problem with the national sample (with a few respondents from many different cable systems) is the diversity of cable systems which comprise that sample. Two-way cable television has been installed in a few franchises, although it has not brought the financial success or the exciting new set of services that was anticipated. In those few systems where upstream amplifiers permit a cable headend to poll responses from converters in subscribers' homes, it is possible for the cable operator to continuously measure viewership. A few studies have been conducted in these systems which have helped clarify viewing patterns (e.g., "Arbitron Survey" 1984). "Peoplemeters," on trial in the U.S. and Europe, link metered household viewing data with viewer's electronic identification of who's watching when.

Cable has been a major threat to the broadcast networks, which expected only to compete with each other, and a problem for national advertisers who want to reach the most people without having them change channels during the commercial. Commercial network audience share has dropped by as much as a third in cable households. Despite the drop in audience share, the cost of advertising time continues to rise. Industry and university studies document an ongoing erosion of network viewing among cable households. Although the audience diversion away from broadcast networks has been substantial, it has not been as large as was initially feared. One Nielsen discovery was that, even though the number of channels viewed does increase as the number of channels available goes up, it does so at a much lower rate. The pay movie cable channels draw the largest audience share (often as much as 10%), but, most of the time, the other cable satellite network stations individually attract very small audience shares.

Another concern, as cable penetration grew, was that viewers, now provided with so many channel options, would watch only the types of programming they liked best. Some diversion from local public television stations has been noted. Cable subscribers tend to watch more different public television stations than nonsubscribers, but for less of their total viewing time than nonsubscribers (Webster & Agostino, 1982). A number of studies examined the impact of cable on news viewing. Three premises were (a) that socially desirable programming (news) would not be watched if sports and movies were available at the same time, (b) that viewers would watch cable news channels instead of broadcast news, and (c) that viewers would watch higher budget newscasts on distant stations carried by cable instead of low budget local news. News viewing levels turned out not to be lower in cable households (Becker, Dunwoody, & Rafaeli, 1983). Viewers who watched cable news such as CNN tended to also watch broadcast news (Collins, Reagan, & Abel, 1983). Diversion from local news to distant stations *did* occur (Hill & Dyer, 1981).

Zapping is a new viewing activity that has accompanied cable, to the dismay of advertisers. Zapping refers to changing channels when a commercial comes on. Before remote control and before multiple cable channels, it was thought sufficient to treat program viewership as a direct indicator of audience size for commercials. Occasional studies suggested that people left the room or tuned out mentally when commercials come on, but not much serious attention was paid. Since the growth of cable, advertisers have begun to request direct measurement of the audience size of each commercial, rather than of the program surrounding the commercial break. One New York-based advertising agency (Market Opinion Research, 1987) predicts that, by the year 2000, 60% of the television viewing audience will not be sticking around for commercials. J. Walter Thompson (1986) has defined three viewing behaviors: flipping, zapping, and zipping. *Flipping* refers to changing channels (flipping around) to some degree rather than watching a show from beginning to end. Thirty-four percent of viewers in a national survey were classified as flippers: 9% were light flippers, 13% medium flippers, and 12% heavy flippers. JWT differentiates zapping (intentionally changing to avoid commercials) from flipping. Only 9% of viewers were classified as Zappers. Then there is *zipping*: fast-forwarding past commercials while watching programs on VCR.

A recent British study reports that one in four VCR viewers watches television commercials while playing back programs taped on a remote control VCR. Users deleted 10% of commercials while watching. Fifty-six percent of the time in all VCR households, viewers fast-forwarded through commercials (Wentz, 1985). A U.S. study shows that 67% of VCR owners skip over ads at least some of the time (Market Opinion Research, 1987). Ogilvy & Mather (1987) rebut the data on flipping, zapping, and zipping, citing their own data from 4 years earlier and concluding, "to say that TV is getting weaker as an ad medium, we believe, is selling ourselves short." Nonetheless, there is talk of producing commercials that people won't want to zap, or commercials that are effective even if they are viewed at high speed. JWT (Thompson, 1986) even suggests showing commercials on part of the screen while the program continues in some way on the rest of the screen.

Cable has been a boon to investors who owned cable systems during its rise in prominence. Between 1970 and 1980, the number of cable systems in the United States jumped 75% (Arbitron, 1981). Cable has become a multi-billion dollar industry, with national penetration nearing 50%. Now, the growth in cable penetration has stopped. Fifty percent penetration is described as a barrier to be broken. Cable systems which once had only to contend with rapid growth now cope with customer satisfaction and high "churn" levels (subscribers dropping pay services or cancelling their subscriptions). A great deal of research seeks ways to predict, encourage, and maintain cable subscribership. Differences between pay cable, basic cable, and nonsubscriber households have been examined, including demographics, amount of viewing, innovativeness, importance of television, use of competing media, and so on (see, e.g., Agostino, 1980; Webster & Agostino, 1982; Katz, 1983; Grotta & Newsom, 1981; Barnes & Kelloway, 1978; Collins et al., 1983; Television Audience Assessment, 1983; Bezzini & Desmond, 1982; Sparkes, 1983a, Sparkes, 1983b).

Cable television continues to affect a variety of information and entertainment industries. But most important for our purposes, cable has changed the television viewing experience for viewers. A pervasive impact of broadcast television has been the set of experiences and acquaintances shared by viewers nationwide. TV has transported us to places we would not otherwise visit (space, China, royal weddings) and introduced us to characters we would not otherwise meet (both real and fictional). The majority of these "horizon-expanding" experiences are collective, shared by a large portion of the American public. The availability of more and specialized channels on cable threatens to reduce the amount and nature of nationally shared media experiences.

Cable also allows viewers to be more active participants in selecting media content. Parker and Dunn (1972) suggest an analogy:

> Broadcast television is like the passenger railroad, taking people to scheduled places at scheduled times. Cable television has the potential of becoming like a highway network, permitting people to use their television sets in the way they use their personal automobile; they may be able to select information, education and entertainment at times and places of their own choosing. (p.176)

Traditional models of mass communication conceptualized viewers as passive receivers to whom sources transmit messages (Heeter, 1987). With cable and other emerging technologies, viewers are being recognized as active selectors of information (Rogers & Chaffee, 1983). Schramm (1983) indicates that we are already beginning to see theories in which the activation agent is the person . . . in place of linear affects models that assume a message is the one-way relationship in which a communicator "does something" to a passive audience.

De Sola Pool (1983, p. 261) predicted a change in research focus that will be brought about by cable and other new technologies. The *effects* of mass media cease to be the only salient question. If people are actively able to choose what they want from an enormous range of information, "this situation makes *the user* more interesting than the effects of the messages on that user." Chen (1984, p. 284) proposes that "passivity and interactivity are qualities of the individuals making use of the media, not the media themselves."

This book consists of 19 chapters which examine the *process of cableviewing*, the *nature of cableviewers*, and especially the *viewing choice process*. The new concept of viewing style is introduced, and individual differences in viewing style are examined. The research presented here represents the change from media effects studies to studies of users and user behavior foreseen by Pool.

The collected studies were conducted between 1982 and 1986 using a variety of research methods, from telephone and mail surveys to door-to-door and in-school questionnaires, to two-way cable measured household viewership. A total of more than 3,000 adults and 1,900 youth, in nine cities from five states, in communities with 36- to 54-channel cable systems, were surveyed.

The book is divided into five sections. In Section One, *The Viewing Choice Process*, a model of program choice is developed and tested. Chapter 2, "The Choice Process Model," defines, operationalizes, and measures the viewing style constructs that are included in studies throughout the book.

In Chapter 3, "A Theoretical Overview of the Program Choice Process," the model is discussed in relation to prior research. Structural differences between cable and broadcast television are examined and effects are postulated.

Chapter 4, "Cableviewing Behaviors: An Electronic Assessment," uses household viewing data collected from a two-way cable system to examine behavioral manifestations of the self-reported viewing style constructs described in Chapters 2 and 3.

Section Two, *Viewing Styles*, examines viewing styles in a number of contexts. Chapter 5, "Profiling the Zappers," is a detailed, multistudy description of zapping and its relationship to other viewing style behaviors and to viewer attributes. Zappers are characterized.

Chapter 6, "New Fall Season Viewing," again uses two-way cable viewing data to examine changes in overall viewing style over time as a new fall season of network programs is introduced. Strings, stretches, and ministretches are again used for comparisons.

Chapter 7, "Watching Saturday Morning Television," compares viewing styles (calculated from two-way cable data) between households with children under 18 and households without children under 18. Channel viewership and viewing style variables are compared.

Chapter 8, "Changes in the Viewing Process Over Time," uses repeated-measure survey data to consider the stability of viewing style between cable and pay cable households, over a 3-month period.

Chapter 9, "Viewing Style Differences Between Radio and Television," takes the television viewing-style variables (channel repertoire, orienting search, reevaluation, and channel familiarity) and applies them to radio listening. Analysis of the survey data address the issue of whether viewing style is consistent across different media.

In Section Three, *Viewer Type, Channel Type, and Viewing Styles*, viewing style is examined from different vantage points. In Chapter 10, "Viewing Context and Style with Electronic Assessment of Viewing Behavior," survey data is combined with two-way cable household viewership data for analysis. Demographic variables are linked with environmental or context variables and with self-reported and computer-monitored viewing style process variables.

Chapter 11, "Parental Influences on Viewing Style," compares parent and child viewing behaviors, both overall and within households. A household viewing style typology is derived.

In Chapter 12, "Sex Differences in Viewing Style," female and male viewing styles are compared across 10 studies.

Section Four, *Program Type and Viewing Style*, considers the relationship between the type of content on a channel and how that channel is watched. Chapter 13, "Channel Types and Viewing Styles," presents an overview of the problem and combines data from several studies to look at 24-hour news channels, pay movie channels, and other specialty content channels. In Chapter 14, "News Viewing Elaborated," two-way cable data is used to precisely document news viewing (from both broadcast and cable only channels) in cable households. Chapter 15, "Playboy Viewing Styles," analyzes how Playboy Channel subscribers watch the Playboy Channel as compared to their viewing style with other channels.

Section Five, *"Subscriber Types,"* identifies differences in viewing style among nonsubscribers, basic cable subscribers, and pay cable subscribers. Chapter 16, "Cable and Noncable Viewing Style Comparisons," develops viewer typologies. Chapter 17 "The Playboy Profile and Other Pay Channel Subscribers," looks at characteristics of subscribers to the different pay channels. In Chapter 18, "Music Video Viewers," regular viewers of MTV are compared to nonviewers in terms of demographics, psychographics, and related music activities.

The last section, *Manipulating Viewing Through Field Experiments*, reports findings from two field experiments which attempted to alter viewing behavior and viewers' likelihood of subscribing to cable. In the experiment reported in Chapter 19, "Free Cable as an Incentive," one free trial month of cable television subscription was offered to all residents in a small town. Chapter 20, "Free System-Specific Cable Guides as an Incentive," involved the design and distribution of a free cable program guide specific to the local franchise area, which was distributed to a randomly selected group of subscribers and nonsubscribers over a 6-month period.

The last chapter summarizes findings from all of these studies and propose a research agenda.

SECTION ONE

The Viewing Choice Process

CHAPTER 2

The Choice Process Model

Carrie Heeter

INTRODUCTION

Most research on selective exposure attempts to link viewing outcome and motives, without consideration of the choice process used by the viewer. Dissonance theorists posit viewer selection of programs consistent with their beliefs and values (e.g., Sears & Freedman, 1974). Arousal theorists posit viewer selection of programs of a type suited to current affective state (e.g., Zillmann, Hezel, & Medoff, 1980). Economic theories of program choice in a multichannel environment assume that viewers have a most-preferred program option that they will select from available options, given the opportunity to do so (e.g., Owen, Beebe, & Manning, 1974). These approaches all implicitly or explicitly assume perfect viewer awareness of program alternatives.

Webster and Wakshlag (1983) integrated many of the divergent program-choice research approaches that have been conceived in the context of broadcast television into a single, broad model. They included viewer awareness in the model, but focus more on viewer availability to watch television as a predictor of the shows or types of shows viewers choose to watch. With only the three major networks to choose from, viewer awareness of program alternatives has not emerged as a particularly significant research issue.[1] Perhaps some viewers are not aware of program alternatives, but becoming so informed would not be particularly difficult.

Across the three commercial networks, the mix of programs or types of programs a viewer can watch varies widely depending on the time of day and the day of the week. Webster and Wakshlag (1983) suggested that, in general, viewers decide first whether or not to watch television at a particular time. Then, if they choose to watch, they select a program. This two-step decision is thought to

---

[1] If one considers public television stations and local independents, the media environment is actually more complex than just three networks. However, studies predicting program choice have examined only network choices, and the vast majority of viewership in noncable households is still devoted exclusively to the networks, except in large cities.

partially explain the weakness of program preferences in predicting viewership. Presumably, if viewers' preferred show types were constantly available, be they violent shows, arousing shows, shows consistent with viewer beliefs, and so on, viewership predictions on the basis of preference would improve.

Cable television alters the media environment in several ways that may influence program choice. Frequently, 35 or more channels are carried,[2] making the task of program selection more complex and making perfect viewer awareness of alternatives more difficult to achieve. Under circumstances of imperfect viewer awareness, viewership predictions may have weaker predictive power. Second, some of the cable channels offer specialized programming of a particular type that is more content consistent (to varying degrees) than the commercial broadcast networks (e.g., Cable News Network or Home Box Office). For viewers who prefer a program type available continuously, their availability to watch television should have less impact on their ability to watch desirable program choices at any given time, and program-type preferences may have stronger predictive power. Third, cable television is frequently accompanied by a remote control channel selector, which may facilitate channel changing as a decision-making tool, encouraging viewers to watch actively (Chapter 5).

Rather than being a fairly closed, narrow-choice situation, selecting programs with cable is potentially an open, complex task. Although the selection process is not necessarily different for cable than it has been for broadcast television, individual differences in decision-making style, such as the ways in which viewers make themselves aware of program options, are likely to have a more substantial impact on viewing patterns and outcomes as the number of channel options increases. This chapter proposes an information-processing model of program choice which takes into account whether and how viewers make themselves aware of program alternatives. Viewer attributes are linked to choice-process variables, and both are used to predict viewing outcomes with cable.

The Choice Process

Program choice, when a viewer does not already know what he or she wants to watch, can be said to have an *indefinite goal*, using Greeno's (1976, p. 479) definition. This type of decision is one where problem solvers "adopt a goal that is indefinite in the sense that it can be satisfied by any of several alternatives, and the selection of an alternative is based on information generated within the process of problem-solving." If we apply his model to program choice, viewers may be assumed to approach a viewing situation with a variety of overt and covert potential goals (or needs) that might be satisfied by watching any number of different available programs. For example, a viewer's affective state may suggest certain immediate needs (e.g., a need for arousal, a need for arousal reduction) in addition to their general content and program preferences. More than one

---

[2] Of all U.S. cable households, 77% receive 22 or more channels; 28% receive 36 or more according to *Cablefile/85* (International Thomson Communications 1985).

program may, to some extent, satisfy different portions of the viewer's needs. Thus, the goal is indefinite, and a single, specific program may not emerge as the fairly optimal choice.

In his book on psychological decision theory, Kozielecki (1981) discussed two dimensions of a decision-making task that can be applied to program choice: complexity and uncertainty. Complexity relates to the number of variables the decision involves. With cable television, there is a limited but large number of program alternatives, and an unknown and perhaps unlimited set of viewer needs, making the task complex. Uncertainty refers to whether the decision maker has full knowledge of a well-defined set of alternatives (a low uncertainty, closed task) or must generate a set of alternatives (a high uncertainty, open task). Although there are a fixed number of potential alternatives (as many as there are programs being broadcast at a given time), viewers do not necessarily consider each alternative. If one equates viewers finding information about a program alternative (through whatever means: channel or guide checking, discussion, etc.) with generating alternatives to consider, then the task is open and uncertain. Different viewers will generate different alternatives to consider.

Greeno's (1976, p. 480) problem-solving approach converges with psychological decision theory in describing a search for information as the first step. He proposed that, "when a problem-solver sets an indefinite goal [e.g., satisfying some combination of needs], the ensuing process often involves a search for information in the situation, without knowledge of how the information will be used." In program choice, the search for information may be an examination of program options (generating alternatives), accompanied by a covert matching of needs with programs that fulfill them (assessing consequences), leading eventually to selection of a matched needs-program option. Kozielecki (1981, p. 260) points out that "a vital role is played in the process [of generating alternatives] by such intellectual abilities as flexibility and originality of thinking." The nature of this "orienting search" for information will be a function of information processing abilities and preferences of the individual viewer, as well as various cable system factors.

Among most individuals, choosing what to watch on television is a decision task that arises on a regular basis. The model proposed here assumes that viewers develop a strategy for program selection that they employ on a regular basis when they are faced with the routinely structured decision-making task of selecting a program. It is important to note that not all television viewing elicits this choice pattern. When viewers turn to television to watch a specific, preselected program, the choice process specified here does not occur.

Orienting Search

With television, two primary sources of viewer information about programs are viewing guides and the channels themselves. Use of a guide may occur independently of channel checking. Both behaviors are manifestations of an orienting search for program alternatives.

Three dimensions of an orienting search are proposed that have implications for viewer awareness of program alternatives: *processing mode* (automatic versus controlled), *search repertoire* (elaborated versus restricted set of channels checked), and *evaluation orientation* (exhaustive versus terminating searches) (see Figure 1).

Shiffrin and Schneider (1977, p. 1) define *controlled processing* as "temporary activation of a sequence of elements that can be set up quickly and easily, but requires attention, is capacity-limited (usually serial in nature), and is controlled by the subject." In contrast, *automatic processing* is "activation of a learned sequence of elements in long-term memory that is initiated by appropriate inputs and then proceeds automatically" (p. 1). Applied to seeking program information, an automatic search might be checking a guide or scanning channels in the order they appear, requiring little mental effort to scan every channel. A controlled search would be channel scanning or guide checking by moving from one selected channel to another selected channel, driven by some selectivity factor(s) other than channel order. *Because automatic searches are not capacity limited and because no selectivity factor is mediating exposure, use of an automatic search strategy is expected to make viewers aware of more different channels than a use of controlled search.*

Regardless of whether channels or channel information are checked in sequence or in selected order, an orienting search can be elaborated including most or all of the channels available on a system, or it can be restricted to a small number of options. By definition, automatic searches lend themselves more to

Figure 1.   Orienting Search Pattern Attributes

I.	PROCESSING	
	a. Automatic*	searching channels in numerical order
	b. Controlled	searching channels in a purposive, regular order other than numerical
II.	SEARCH REPERTOIRE	
	a. Elaborated*	a search pattern which includes all or most channels
	b. Restricted	a search pattern which includes a limited number of channels
III.	EVALUATION	
	a. Exhaustive*	searching *all* channels of an individual's search repertoire and returning to the best option
	b. Terminating	searching channels of an individual's search repertoire in the viewer's usual order, only until the first acceptable option is located

*indicates orienting search attributes most likely to lead to full awareness of program alternatives.

elaborated searches. *An elaborated search is expected to result in greater viewer awareness of program options.*

Shiffrin and Schneider (1977) also identify different *outcomes* sought in the choice evaluation process. In a *terminating search*, the viewer abandons the search when the first option that meets some minimal standard is located (in economic terms, a "satisficing" approach). An *exhaustive search* includes scanning of *all channels that compose the individual's search repertoire*, followed by a return to the *best* alternative. Thus, for an automatic, elaborated searcher to also search exhaustively, the viewer must check every channel on the system. For a restricted and/or controlled search, the viewer must check every channel within that set of channels usually checked, in the usual order.

For each channel alternative checked, a viewer must make a judgement. There are three possible outcomes of the alternative evaluation: reject, accept, or consider and check other options. A terminating search involves only accepting or rejecting a program alternative. Short-term memory is believed to be the locus of decision making, and decision processes are therefore strongly codetermined by memory characteristics (Kozielecki, 1981). There is controversy over whether the limitation of the short-term memory capacity is 5–9 portions of information or 3–5, but agreement that the capacity is limited and varies across individuals. An exhaustive search requires greater memory capacity, storing viable program options in short-term memory, and exposes viewers to more channels more often. *An exhaustive search pattern is expected to result in greater viewer awareness of program and channel options, and therefore, use of more different channels.*

Reevaluation

Choosing what to watch is a recurring issue. Once a program has been selected, the search process may end for the program's duration, at which point it begins again. For some viewers, and particularly for viewers with remote channel selectors, frequent or even continual *reevaluation* of that initial choice may occur, while the selected program is on. An obvious period for reevaluation is at commercial breaks during a show. With cable, Television Audience Assessment, Inc. (1983) reports that 39% of cable subscribers almost always or often change channels during commercials. Nielsen (1983a) identifies twice as many tune-outs occurring during commercial minutes as during noncommercial minutes. Those who zap commercials are exposed to more channel options (while changing) than viewers who watch commercials or leave the room.

More critical viewers may even reevaluate program choice during a show, other than at commercials. A special case of this is watching more than one show at a time by moving back and forth between channels. Chapter 16 identifies multiple show viewing in one-third of cable households. *Zappers of programs and commercials alike should be more familiar with what channels are available, and should watch more different channels.*

Channel Familiarity

A methodological study by Arbitron (1983, p. 13) revealed that "cable subscribers are not aware of all the services available to them, or even of what service they are viewing at any given time." Familiarity with the system may vary widely among viewers. It is one means of assessing awareness of cable programming. Channel familiarity is expected in part to be a function of orienting search and reevaluation patterns, as well as other individual variables.

Channel Repertoire

Numerous outcomes of a choice process model can be measured. Typically in broadcast models, the outcome considered has been viewership of a specific program or program type. The impact of cable television (on broadcast television) is often reported in terms of differences in viewership of channel types (e.g., Webster, 1983, comparing viewership of network channels, independents, PBS, pay movie, and basic cable).

These kinds of aggregated comparisons do not address the impact of cable on individual viewers. Cable viewers watch more different channels on a regular basis than broadcast viewers, but they watch far fewer channels than the total number available; an average of about 10 different channels are watched at least once a week (Nielsen, 1983b; see Chapter 4, where Heeter et al. introduce the concept of *channel repertoire*—the set of channels watched regularly by an individual or household. Channel repertoire can be measured simply as the number of different channels watched. It can also be differentiated, to identify the number of cable-only channels watched. In this study, channel repertoire will be used as an outcome variable, *expected to be influenced by choice process patterns and channel familiarity.*)

Figure 2 summarizes the choice model to be tested. The first stage of the choice process is an orienting search of some kind, which may involve guide use and/or channel checking. Guide checking and channel checking may occur independently. Viewers may rely exclusively on either program information source, both, or neither. However, use of either or both sources should increase channel familiarity and channel repertoire. The three modes of orienting search hypothesized to result in maximal awareness (automatic, elaborative, and exhaustive) are expected to be positively correlated with each other. Once an initial program is selected, it may be watched in its entirety, or the viewer may engage in reevaluation of the choice before the program ends. Here also, guide use or channel changing could occur. At some point, the selection process begins again. The extent to which, and the manner in which, viewers engage in active selection behaviors are expected to have an impact on their general awareness of cable channels (channel familiarity) and on their channel repertoire (set of channels watched regularly).

Viewer Attributes

In addition to developing measurement systems and testing the choice model internally, certain viewer attributes are expected to relate to choice process behaviors. Specifically, age, sex, education, pay cable subscribership, and novelty-seeking were hypothesized to relate to channel choice activity. Figure 2 diagrams these predictions.

Krull and Watt (1975) suggest a curvilinear relationship between age and information processing abilities, with the highest abilities found for individuals in their twenties, and lesser ability for those older and younger. Because exhaus-

Figure 2.   Choice process model (predictions).

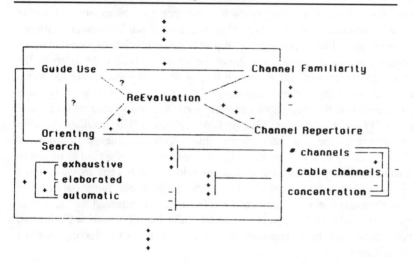

Trait	Guides	Or. Search	Reevaluation
Age (younger)	?	–	–
Sex (male)	?	+	+
Education	+	+	+
Pay Subscription	+	+	+
Novelty-Seeking			
Cognitive	+	+	+
Sensate	?	+	+

tive, elaborated, and automatic orienting searches and frequent reevaluation would tend to require more information processing, age is expected to be negatively correlated with guide use, orienting searches, and reevaluation among adults.

The proposition that males will approach the choice process differently from females derives in part from documented differences in math and spatial abilities between males and females (e.g., Chen & Paisley, 1983). Observational experiments with public-access teletext report that males more frequently used the public terminals, and that the sexes appeared to approach the new devices differently. *Males are expected to be more comfortable with the remote control device and therefore to engage in more orienting searches and reevaluation.*

Education was expected to be at least moderately correlated with information processing abilities, in part because those skills should be enhanced by training and application. *Highly educated respondents were expected to be more discriminating and selective viewers, and to make themselves aware of program alternatives.*

Prior studies have found that pay cable subscribers use guides more often than basic cable subscribers (62% versus 55% of the time; see Television Audience Assessment, Inc., 1983), presumably in part because special pay service guides are normally provided to them, and in part because they want to be aware of the programming which they value enough to pay an extra subscription fee to receive. Analysis of cable viewership using data collected over a two-way cable system found that the pay cable channels were the most frequently checked cable-only channels (Chapter 13). Perhaps because of the availability of a desired program type, pay subscribers more frequently engage in orienting searches. In addition, a curious viewing phenomenon was observed in regard to pay channel viewing: It was rare for a household to watch a pay movie channel continuously for an entire movie. Viewers either tuned in late, tuned out early, or changed channels for some amount of time and then returned to the movie (Chapter 13). Nielsen (1983a) reports that pay cable households show the highest level of channel switching. This pattern is expected to apply to orienting searches and reevaluation.

Zuckerman (1979, p. 315) identified and defined sensation-seeking as a personality trait—specifically, "every individual has characteristic levels of stimulation and arousal for cognitive activity, motor activity, and positive affective tone." Maddi (1968) identified a similar construct—"variety-seeking" as a gratification sought from media. Jeffres (1978) attempted to differentiate "media-seeking" and "content-seeking" with cable. An important implication of this distinction is to point out that viewers may on occasion be seeking arousal (limbic or cortical) rather than looking for a show to watch. With cable's greater variety of channels, it may provide greater content-free stimulation. The study of two-way cable data found some households that consistently devoted the majority of their viewing day to channel scanning. At the extreme, one household spent a total of more than 4 hours changing channels at least every 4 minutes in a single day. This extreme example reflects ongoing orienting searches and continual reevaluation.

Kozielecki (1981, p. 299) related demand for stimulation to decision-making style. "For each individual, there is an optimal level of stimulation that ensures satisfaction. Whenever the actual stimulation falls below a critical level the person starts searching for additional stimuli and situations, i.e., undertakes new organizational or intellectual tasks. Whenever stimulation exceeds a critical level, the person seeks to reduce it by say, withdrawing from some activities." He found that 60% of decision makers with a high demand for stimulation preferred more risky (probabilistic) decisions, whereas 61% of those with a low demand for stimulation preferred low-risk (deterministic) decisions.

Donohue, Palmgren, and Duncan (1980) propose that individuals enter "information-exposure situations" (e.g., watching cable) with the expectation of achieving or maintaining that optimal stimulation state. *A high demand for arousal is expected to be associated with more elaborate orienting searches and more reevaluation.*

METHODS

A door-to-door survey was conducted by trained student interviewers in a medium-sized midwestern city served by a cable system franchised to Continental Cablevision. The 35-channel system carried two affiliates from each commercial network, two public television stations, two pay channels (HBO and Cinemax), 11 satellite-delivered nonpay cable services, 11 local and/or state government, education, and public-access channels, one superstation, and two independents.

A total of 32 cluster areas were randomly selected from detailed city-planning maps, evenly distributed across city limits. A random start point and path were identified within each cluster, and a skip pattern of every fifth house was used. Only households that subscribed to cable were interviewed. Of the homes contacted, 67% received cable. (A telephone survey 1 year earlier found 62% penetration.) There was a 7% refusal rate. Less than 1% of the interviews were terminated in progress. Of cable households contacted, 15% indicated willingness to be interviewed at some other time, but were not reached in a follow-up attempt. Interviews lasting 20–30 minutes each were completed in 232 cable households.

The sample was 47% male. The average age was 40. A total of 15% had not completed high school, 38% were high school graduates, 32% had attended some college, and 15% had earned at least a bachelor's degree. Average range of annual household income was $20,000–$30,000. There were three residents, one of them a child under 18, in the average household.

The questionnaire included items on:

   Viewer Attributes
   Guide Use
   Orienting Search
   Reevaluation
   Channel Familiarity
   Channel Repertoire

as well as some television viewing control variables. Item examples appear in subsequent parts of the methods section. Interrelationships among the choice process constructs were examined using simple correlations. Multiple regression was used to test the research hypotheses.

Viewer Attributes

Sex, age, income, number of children, and pay cable subscribership were described in methods. Pearson's (1970) Novelty-Seeking Scale (NES) was used to assess demand for stimulation. The first 20 items from NES (evenly distributed across the external sensate and cognitive novelty seeking) were included in the door-to-door survey for this study. When factor analyzed using varimax orthogonal rotation and constrained to two factors, the expected factor types emerged. External cognitive and external sensate indices approached normal distributions and were uncorrelated. Eight items reflected external-sensate (e.g., liking or disliking being on a raft in the middle of the Colorado River; being at the top of a roller coaster ready to go down), and six represented external-cognitive novelty seeking (e.g., reading the *World Almanac*; figuring out how a light meter works). Additive indices of the items loading .4 or higher on the appropriate factor were constructed. Cronbach's alpha for external-cognitive was .68, and for external-sensate, .70.

Control Variables

Typical weekday television viewing time was assessed by adding self-reports of the amount of viewing in morning, afternoon, and evening. The mean was 5.2 hours.

   Because program choice was being modeled, respondents' dominance over the channel selector and participation in group decisions was assessed. About one-fifth of respondents rarely made decisions all by themselves, but 52% perceived themselves to be involved in the decisions at least 3/4 of the time.

   Because the choice process proposed here does not apply to situations where viewers already know what they want to watch, respondents were asked how much of the time they watched shows other than their regular daily and weekly shows: 18% watch regular shows almost exclusively; 29% do so three-fourths of the time, one-fifth about half the time, 11% only one-fourth of the time, and 21% almost never watch a show regularly.

In this cable system, three channel selectors were available: (a) a remote, hand-held, digital selector where channels are accessed by channel number; (b) a slide selector with remote extension cord, where a small knob is moved across all channels in order to get from one to another; and (c) push-button selectors, with 12 buttons in a row and a 3-position switch on the side to yield 36 channel positions. Unfortunately, there was an extra charge for remote selectors, and push-button devices did not come with an extension-cord option. Type of channel selector is thus confounded with remote control. People with remote selectors had them because they were willing to pay for the convenience. Because of the confounding, remote control was used as a control variable, rather than including it in the hypotheses: 34% had digital, 22% slide, and 43% push-button (nonremote) selectors.

Guide Checking

Guide use questions asked how much of the time respondents checked a guide before turning the set on, and how much of the time they checked a guide while watching television. Response categories were: almost all the time, 3/4 of the time, 1/2 of the time, 1/4 of the time, and almost never. (These response categories were used for numerous items throughout the questionnaire.) Guide checking before viewing was bimodal, with 30% almost never and 35% almost always doing so; 42% almost never refer to a guide while they are watching tv, and one forth check a guide 3/4 to almost all of the times the set is on. In addition, respondents were asked whether they would turn the set on or check a guide before watching TV. The three guide-use items were combined into a single additive index, with a standardized item alpha of .70. The orienting search distinctions used with channel checking should also apply to guide checking, but this measurement was omitted.

Channel Familiarity

The measurement of channel familiarity was a principal motivation for door-to-door interviews. Respondents were handed a blank grid of channel numbers representing their cable system and were asked to either fill in or instruct the interviewer what to fill in as call letters, originating city, general content, or any other identification of all the channels they were familiar with on their system. Space was also provided for channels whose content but not location on the system was known. Spaces were numbered from Channels 2 through 37; no respondent identified content for the nonexistent 37. Extensive pretesting identified a distinct lack of awareness of cable channel numbers beyond Channel 13 (the last of the

12 buttons on the first tier) among pushbutton selector households. The more industrious pretest subjects used their fingers to try to match their spatial memory of where the channel was with the channel number to which it should correspond. A special 3-row grid was developed and used for pushbutton respondents, alleviating that problem.

On the average, 9.4 channels were correctly identified by content and location, .8 were known but not by location, and .8 were incorrectly identified. Of respondents, 23% were able to identify only 0–3 channels. A common response reported by interviewers of those respondents was "I don't know, I just watch TV," and, even with seemingly sincere effort on their part, they had little knowledge of what was available over cable, let alone the broadcast channels. At the other extreme, 23% named 14–27 channels.

Orienting Search

Identification of viewer's search patterns was the other major impetus for door-to-door surveys. Holding the grid respondents had used to report channel familiarity, interviewers asked them to imagine a situation in which they wanted to watch television, did not know what was on, and no guide was available. They were asked what channel they would turn to first. With repeated prompting ("then what . . . then what . . . ?") the interviewer elicited the respondent's search patterns, noting which channels were checked and in what order. Search pattern identification did not seem to be restricted by channel familiarity already marked on the grid. One-third of respondents reported checking fewer channels than those with which they were familiar. Two-thirds checked more channels than they could identify by name or content.

A count of the total number of channels checked was used as a measure of how elaborated or restricted a viewer's orienting search was. Two very distinct approaches emerged: 42% of respondents reported checking 34 or 35 channels, while 55% checked between 0 and 14 channels (averaging 5.9 among the restricted search group). Three percent were located between those extremes.

The number of channels checked in sequential order was used as a measure of whether the search pattern tended to be automatic (more channels checked sequentially) or controlled (channels checked out of sequence). Channel 4–then 5–then 6 would be two sequential checks. Channels 5–then 4–then 6 would be one sequential check; 37% checked between 27 and 35 channels in order (their mode was 35, encompassing 20% of all respondents); 53% checked between 0 and 5 channels sequentially (their mode was 2, encompassing 25% of respondents).

Respondents were asked whether, during their channel searches, they stopped at the first show that looked good (a terminating search) or finished checking their usual channels and went back (an exhaustive search). Half reported terminating searches, 46% used exhaustive searches, and 4% didn't know.

To begin to assess the validity of the assumption that an orienting search pattern represents what viewers regularly do, respondents were asked how much of the time they used the order and pattern they had specified: 53% claimed to do so almost all the time, 10% three fourths of the time, 13% half the time, and 12% each one quarter and almost none of the time, lending credence to the assumption that regular, patterned behavior occurs.

Automatic and elaborated searches were correlated .97. Dividing respondents into three groups (automatic, controlled, and in-between) on the basis of number of channels checked sequentially, those engaging in controlled searches had an average restricted search pattern of 5 channels. Those engaging in automatic searches checked an average of 35 channels, in an elaborated search. Those in the mixed group checked 25 channels (ANOVA, p = .0000). Viewers engaging in terminating searches tended to use a controlled sequence and checked fewer channels (an average of 16) compared to exhaustive searchers (average = 22, t-test, p = .002). Terminating searchers used their search pattern less frequently (about 66% of the time) than exhaustive searchers (77%, t-test p = .024).

Reevaluation

A total of seven items assessed channel changing behaviors. Of them, three (frequency of changing between shows, changing during a show at commercials, and changing during a show other than at commercials) were intended to comprise a Guttman scale, ranging from least to most extreme behavior. One-third of respondents almost never changed between shows; 55% almost never changed at commercials, during a show; and 69% almost never changed in the middle of a show, other than at commercials. The items were subjected to Guttman scale analysis with a prespecified order. The initial coefficient of reproducibility was .93, and the coefficient of scalability .82. Reclassification of the 13 cases which indicated channel changing during commercials and/or during a show, but no changing between shows, to include changing between shows improved the coefficient of reproducibility to .96 and scalability to .87. Distribution of subjects from least to most channel changing included 27% in group 1 (nonchangers), 21% in group 2, 27% in group 3, and 24% in group 4 (most frequent changers).

Table 1 presents mean comparisons from oneway ANOVAs of the Guttman scale groups along the other four channels changing measures, to profile viewer types. At one extreme is the group which almost never changes channels, averaging perhaps one change every hour and a half during regular shows (shows watched almost every day or week), and one change every 45 minutes during nonregular shows. They search for something to watch less than twice a day, and only 9% of their members ever watch more than one show at a time. On the other extreme is the group which changes channels three out of every four commercial breaks, in addition to changing channels two of every three periods between shows, in addition to changing channels in the middle of 44% of the shows they do watch. They change channels 9 or 10 times an hour during regular shows, 12 times an hour during nonregular shows, and search for something to watch at least three times a day. Nearly two-thirds of them sometimes watch more than one show at a time. All differences were significant at p = .0000. The Guttman scale was therefore deemed an appropriate measure to reflect reevaluation.

Channel Repertoire

Three measures of channel repertoire were used in this analysis (although many other representations are possible.) Using the same grid filled out for orienting search pattern and channel familiarity, interviewers asked how many days per week the respondent typically found something to watch on each of the channels; 40% of respondents viewed more channels regularly than they could identify with channel information. One created variable was the number of different channels watched 1 or more days per week (essentially, a "composite cume" measure). A second index counted regular viewership of channels unique to cable. An index of concentration of viewership across the different channels regularly watched was also computed.

The average number of channels watched regularly was 7.6. Ten percent identified two or fewer channels that they watched at least once a week; 11% watched 12 or more channels. Overall, estimates using this method were lower than the average of 10 found by some other studies. In part this may be due to having 11 access channels (where large audiences would not be expected) of the 35 available. In part it may be measurement technique. Cable channels consti-

Table 1.   Reevaluation Measures: Interrelationships*

Items Constituting Scale:	Guttman Scale Groups			
	0	1	2	3
Percentage who change between shows	0	56	51	64
Percentage who change during a show at commercials	0	0	55	73
Percentage who change during a show, not at commercials	0	0	9	44
Related items:				
Times per half hour change while watching regular shows	.39	.32	1.5	4.8
Times per half hour change during nonregular shows	.66	.81	3.1	5.9
Times per day search for something to watch	1.7	1.7	2.9	3.3
Percentage who watch more than one show at a time	9	23	33	6
n	64	48	62	56

*Overall differences were significant at p < .0001.

tuted 2.7 of the 7.6 channels viewed. Number of channels and cable channels watched were strongly correlated ($r = .91$, $p = .000$).

The viewing concentration index used was the Herfindal index from economics, used by the federal government as evidence in antitrust cases (Scherer, 1970, p. 51). Time allocated to each channel was divided by the total time allocated across all channels. That quantity was squared, and the collection of squared scores summed to form the index. Herfindal indexes range from 0 (perfectly even distribution of viewership across an infinite number of channels) to 1 (a "perfect monopoly," with all viewing allocated to a single channel). The index is a function of both the number of competitors (different channels receiving any viewership) and proportion of the marketplace (total viewing) allocated to each channel. The average concentration was .24. As an example of how the index varies, a one-way ANOVA of concentration by the Guttman reevaluation scale was calculated. Means from least to most channel changing groups were .31, .39, .21, and .18 ($p = .002$), indicating that frequent channel changing is significantly associated with lower viewing concentration. Number of cable channels viewed and overall number of channels viewed were significantly correlated with the concentration index ($r = -.61$, $r = -.72$, respectively).

RESULTS

Model Interrelationships

Figure 3 presents the choice model with simple correlations for each variable combination. When we examine the interrelationships among the model components, guide use is uncorrelated with either of the channel changing variables (orienting searches or reevaluation) and, counter to hypotheses, also independent of the three measures of channel repertoire. Guide use was significantly, but weakly, associated with channel familiarity ($r = .12$).

All of the other model interrelationships were significant, in the expected direction. For orienting searches, exhaustive, elaborated, and automatic searches were each associated with more reevaluation, greater channel familiarity, and larger, less-concentrated channel repertoires. The level of association generally ranged around .20 to .25. Automatic searching showed the lowest correlations, and also was the most questionable in terms of variable measurement. Reevaluation was somewhat more strongly related to channel familiarity and channel repertoire, averaging about .30. Channel familiarity was strongly correlated with channel repertoire, at .59 for total number of channels viewed, .56 for number of cable channels, and $-.42$ for viewing concentration.

Model Testing

Table 2 presents simple correlations between the independent variables and the model variables being predicted. Table 3 contains standardized beta weights for

Figure 3.   Choice process model (correlations). Correlations of .10 or higher were significant at p < .05.

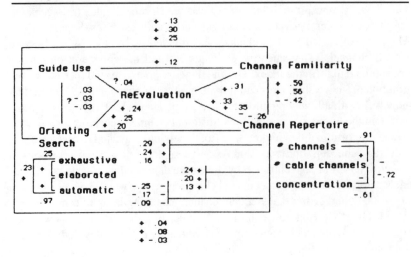

Trait	Guides	Or. Search			Reevaluation
Age (younger)	? -.11	- -.27	-.23	-.27	- -.39
Sex (male)	? -.13	+ .02	.13	.14	+ .30
Education	+ .03	+ .03	.06	.07	+ .10
Pay Subscription	+ .21	+ .13	.02	.05	+ .17
Novelty-Seeking					
Cognitive	+ .14	+ .16	.02	.05	+ .16
Sensate	? -.01	+ .17	.17	.18	+ .34

the entire set of predictor variables in each multiple regression equation, the multiple R, and significance levels.

As hypothesized, pay cable subscribership and cognitive novelty-seeking were positively correlated with guide use (r = .21, r = .14). Contrary to expectations, education did not relate to guide use. Younger females, with more children, also tended to use guides more. When the entire set of independent variables is controlled for, only sex (female) and number of children emerge as significant predictors, with a significant multiple correlation of .35.

Males more often had elaborated, automatic orienting searches, but sex was not correlated with exhaustive versus terminating search. Age showed the strongest correlation, with younger viewers engaging more in the predicted behaviors. Contrary to expectations, education was unrelated to all but one orienting search measure—automatic versus controlled—and that relationship was

Table 2. Correlation Matrix

	Orienting Search					Channel Familiarity	Channel Repertoire		
	Guide Use	Exhaustive	Automatic	Elaborated	Reevaluation		Channels Viewed	Cable Channels	Concentration
Demographics									
sex (female)	.13*	−.02	−.13*	−.14*	−.30*	−.22*	−.11*	−.17*	.13*
age	−.11*	−.27*	−.23*	−.27*	−.39*	−.23*	−.21*	−.28*	.18*
income	−.04	.04	.06	.07	.18*	.16*	.14*	.12*	−.08
education	.03	.06	−.13*	−.09	.10	.18*	.16*	.13*	−.18*
number of children	.16	.09	−.05	−.04	−.09	.02	.04	.08	−.04
Cable subscription									
pay subscriber	.21*	.13	.02	.05	.17*	.05	.15*	.27*	−.10
remote control	−.09	−.03	−.06	−.06	.18*	−.03	.07	.10	.02
Novelty-seeking									
sensate	−.01	.17*	.17*	.18*	.34	.20	.16	.22	−.15*
cognitive	.14*	.16*	.02	.05	.16*	.13*	.16*	.18*	−.23*
TV viewing									
TV hours	−.04	−.08	−.03	−.00	−.01	.01	.18*	.11*	−.17*
habitual viewing	−.01	−.10	−.04	−.02	−.20*	−.11*	.04	.00	−.06
choice dominance	−.01	.01	.00	.01	.04	.09	.13*	.13*	−.03

*Indicates significance at $p < .05$.

Table 3. Multiple Regression Matrix

	Orienting Search					Channel Repertoire			
	Guide Use	Exhaustive	Automatic	Elaborated	Reevaluation	Channel Familiarity	Channels Viewed	Cable Channels	Concentration
Demographics									
sex (female)	.17*	.01	-.09	-.09	.15*	-.13	-.06	-.12	.13
age	-.04	-.38*	-.26*	-.30*	-.26*	-.13	-.01	-.00	.16
income	-.04	-.06	.09	.08	.01	.01	-.02	-.04	.05
education	.08	.04	-.18*	-.13	-.01	.19*	.19*	.14*	-.25*
number of children	.19*	-.03	-.12	-.12	-.17*	-.02	.01	.05	.02
Cable subscription									
pay subscriber	.14	.09	-.06	-.04	.08	-.04	.09	.20*	-.05
remote control	-.07	-.13	-.12	-.13	.10	-.06	.10	.10	.04
Novelty-seeking									
sensate	.01	-.08	-.04	-.06	.03	-.05	-.05	.03	.13
cognitive	.12	.08	.00	.04	.09	.06	.13	.11	-.21*
TV viewing									
TV hours	.01	-.02	-.07	-.03	.08	.06	.24*	.17*	-.25*
habitual viewing	-.03	-.06	.01	.05	-.20*	.04	.06	.08	-.10
choice dominance	.00	.01	.03	.04	.06	.09	.15*	.14*	-.03
Orienting search									
guide use					-.05	.15*	.08	.08	-.02
exhaustive					.12	-.02	.18*	.12	-.15*
automatic					xxxxx	xxxxx	xxxxx	xxxxx	xxxxx
elaborated					.11	.25*	.18*	.12	-.10
Reevaluation						16*	.20*	.21*	-.11
Multiple R	.35	.38	.30	.30	.57	.49	.58	.59	.52
Significance	.024	.005	.149	.153	.000	.000	.000	.000	.000
Mult R w/Chan Fam							.75	.74	.59

*Indicates beta significance at $p < .05$.

28

negative. Sensate novelty seeking was related to the predicted orienting search pattern. Cognitive novelty seeking and pay cable subscribership were both associated with exhaustive rather than terminating searches, but not correlated with the other search measures. Of the three multiple regressions, only the exhaustive-terminating was significant (questionable also because it was a dichotomous dependent variable). Only age and, for automatic versus controlled, education emerged as significant predictors.

For reevaluation, all predicted associations except education emerged as significant: male, younger, pay subscribing, sensate and cognitive novelty seekers engaged in more reevaluation. Availability of a remote control selector and less habitual show viewing were also related to more reevaluation. Age, sex, number of children, and amount of habitual viewing all emerged as significant in the regression, with a multiple R of .57. (Because automatic and elaborated searches were correlated .97, only elaborated was included in the regressions, to avoid multicollinearity.)

Channel familiarity was related to being male, younger, of higher income and education, seeking more novelty, and viewing less habitually. In the regression, the three hypothesized choice model variables (guide use, number of channels checked, and reevaluation), plus education, were the significant predictors, overriding the other demographic variables and yielding a multiple correlation of .49.

Channel repertoire was related to being male, younger, of higher education and income, subscribing to pay cable, cognitive and sensate novelty seeking, watching more TV, and dominating the choice process. However, education, pay subscribership, and choice dominance were not significantly correlated with viewing concentration. In the regressions, significant predictors were primarily the choice process and television-viewing control variables. Education did predict more channels viewed and lower concentration. Pay cable subscribership predicted number of cable channels watched (but not overall number of channels watched). Television-viewing time was a predictor for all the channel repertoire measures, and choice dominance predicted number of channels (cable and overall) viewed. Having an exhaustive orienting search predicted number of channels watched and concentration. Having an elaborated search and engaging in reevaluation predicted number of channels watched and number of cable channels watched, but not concentration. The multiple correlations were .58 and .59 for channels and cable channels, and .52 for concentration. When channel familiarity is included as a predictor, the multiple correlations jump to .75, .74, and .59, respectively.

CONCLUSIONS

This chapter sought to define and operationalize constructs relevant to program choice and viewer awareness, to propose a model and examine interrelationships, to link these constructs with other viewer attributes, and to test the model, controlling for TV viewing and demographic variables.

Respondents were able to describe their behavior in terms of different orienting search processes, and the majority claimed that this description referred to a pattern they used routinely. Cable television program choice does appear to meet the conditions of a complex, uncertain, routine decision-making task. Perhaps it is an appropriate situation for further study of decision making.

Elaborated and automatic searching were correlated .97. Although this is consistent with the expectation that the two behaviors co-occur, it also raises questions about whether they reflect separable approaches or the same approach, and also, whether automatic versus controlled searching was measured adequately. As expected, exhaustive and elaborated patterns are somewhat related. But in the regressions, they significantly predict different dependent variables.

Reevaluation and orienting search as measured in this study are both channel changing behaviors. They too are somewhat correlated, but predict different dependent variables. For all seven channel-changing variables, respondents are consistently differentiated, suggesting that channel-changing orientations are viewer characteristics worth examining further. Zapping has been a growing concern in the advertising industry. This study suggests that zapping predominates among young males, and that it is a specific type of viewer who does not watch commercials.

Channel familiarity was predicted by guide use, orienting search, reevaluation, and education, controlling for other demographics and TV viewing. The basic tenent of the choice process model proposed here is quite strongly supported—that channel changing and guide use increase viewer awareness of channel options. Thus, it appears to be in the cable industry's best interest to encourage channel changing and guide use in hopes of increasing awareness of cable channels. Current awareness levels of what is available over cable are very low, as demonstrated by baseline levels of channel familiarity.

The three operationalizations of channel repertoire are highly correlated (from at least .61 up to .91). They are predicted by somewhat different combinations of variables. For example, pay cable subscribership predicts only number of cable channels watched, not total number of channels watched. Perhaps among nonpay subscribers, a different cable-only channel takes the place of a pay channel. Cognitive novelty seeking significantly predicts only a low concentration of viewership.

The strong relationship between channel familiarity and channel repertoire merits further examination. It is likely that some mutual causality operates between them. However, the possibility of increasing channel repertoire by manipulating viewers' familiarity with cable channels should be explored. Again, this should be of interest to the cable industry. Those who change channels and check guides must also know the most about cable and watch more different channels. And that set of viewers can be differentiated by demographics: Specifically, they tend to be young adult males. Perhaps more attention should be paid to channel advertisement and distribution of complete program guides.

The model delineates general relationships between its components, most of which are supported by the data. Multivariate analysis finds that the expected choice-process variables do explain significant amounts of variance in channel familiarity and channel repertoire, above and beyond what viewer demographics and viewing habits account for. This is fairly strong support suggesting that choice-process patterns influence awareness. The model was not tested in full, using path analysis. It was deemed more appropriate for a first test of a model to specify and test general hypothesized relationships, rather than test the precise and complex set of interrelationships simultaneously. At some point, path analysis should be conducted.

The model is tested using in-person interviews to elicit role-playing responses ("let's say you're watching TV and you don't know what is on . . ."). More accurate measures of behavior, either by two-way cable monitoring of actual viewership and channel changing, or by some technique such as "think aloud" descriptions of choice while subjects select programs, could better validate the constructs.

The consistent relationships between many of the viewer attributes and the choice process variables is worth noting. Males and females approach program selection quite differently, with females checking a guide more, and males changing channels (at all times) more, being familiar with and watching more different channels, and engaging in less concentrated channel use. The gender differences remain significant only for guide use and reevaluation regressions.

Younger adults engaged in more of all the behaviors expected to result in diverse and extensive channel awareness and use. Controlling for other variables, age remained a significant predictor only of the channel changing variables. Education was more related to outcome than process variables, significantly predicting channel familiarity and repertoire. In fact, counter to hypotheses, education was a negative predictor of number of different channels checked in orienting searches. Perhaps those with the greatest information-processing skills check channels more selectively. Although demand for stimulation was correlated with most of the choice-process variables, only cognitive novelty seeking remained significant controlling for other variables in predicting viewership concentration.

Pay cable subscribership significantly predicted only number of cable-only channels watched. Many other cable studies have sought to differentiate pay cable from basic cable subscribers. On the basis of these data at least, that distinction does not appear to be very important.

Presence of a remote control channel selector would logically be related to amount of channel changing. Remote control was confounded with channel selector type in the cable system studied. With that caveat, note that it was correlated only with reevaluation. Further research should clarify the role of channel selector type and remote control in the choice process.

It is also possible that a small set of viewer-types, defined by different choice process approaches, could be developed. This would permit integration of the choice process model with other television use and impact research. Selective exposure and attention research are prime candidates for this integration. Viewer awareness of program alternatives could be controlled for in predictions of viewership.

## CHAPTER 3

A Theoretical Overview of the Program Choice Process

Carrie Heeter and Bradley S. Greenberg

INTRODUCTION

The most pervasive assumption of research on program choice has been that, when viewers select a program to watch, they *evaluate all program options available* at the time, and *select the one which best fits* some criterion. In a television environment where only three networks are available, this assumption rarely has been questioned. However, in cable television environments, as the number of program options increases vastly, that assumption becomes less plausible. One implication of the cable environment for selective exposure research is to suggest that the actual choice process needs better articulation.

Many different approaches, both theoretical and atheoretical, have been advanced to predict program choice. Most of them move directly from predictor factors to choice outcome, on the basis of a program choice maximizing assumption. Articulation of a choice process could help identify conditions under which a predicted outcome would be most likely to occur. For example, a number of studies have found a relationship between aggressive predisposition and the amount of television violence viewed (e.g., Stein & Friedrich, 1972; Atkin, Greenberg, Korzenny, & McDermott, 1979; Lefkowitz, Eron, Walder, & Huesmann, 1981). Subjects whose process of choice does involve evaluation of all alternatives and selection of the best program will be aware of all violent and nonviolent program options. Those who check only a few options may not be aware that nonviolent (or violent) program options exist, and therefore their program selection might not be the program they most prefer. Thus, subjects who evaluate all program alternatives should demonstrate the strongest relationship between aggressive predisposition and exposure to violent shows. Similarly, Zillmann and colleagues are finding evidence for a relationship between affective state and program choice (e.g., Zillmann, 1982; Zillmann, Hezel, & Medoff, 1980). When viewers turn to television to relax or unwind, they are expected to seek absorbing content which can serve to disrupt their internal rehearsal of aversive situations that motivated them to seek relaxation, and which will not remind them of these situations. Whereas, when viewers watch for excitement, they seek arousing fare. Here again, those viewers who use a maximizing choice process should best demonstrate this relationship. In contrast, those who do not screen the full set of available options may have their needs less satisfied.

Marketing research models of program choice also ignore the choice process in their attempts to predict viewership patterns from program-type preferences. Program-type preferences have been defined, measured, and used to differentiate consistently among different subgroups of viewers (Gensch & Ranganathan, 1974). However, attempts to predict viewing from preferences and liking of shows have met with little success. Scheduling factors (such as program type availability at any given time, program length, channel, and day of week: e.g., Bowman & Farley, 1972; Bruno, 1973, Frank, Becknell, & Cloaskey, 1971) as well as viewer availability to watch TV when preferred show types are on (Gensch & Shaman, 1980) appear to confound any observed relationships between preferences and viewership. Scheduling factors confound program preference models in broadcast television environments. Cable television may improve prediction by offering a more consistent and larger array of program options.

Webster and Wakshlag (1983) integrate many of the divergent program choice research approaches into a single "model of program choice." They identify the following components: viewer availability, viewer awareness of program options, program and program type preferences, viewer needs, viewing group, and the structure of available programming. The approaches they bring together were conceived of, and studied in, the context of network-dominated broadcast television. The present chapter will use Webster and Wakshlag's framework to detail changes and additions to those components when the television environment is cable television instead of broadcast. A model of the actual choice process with cable will be advanced which may also have utility in broadcast situations. Research on cable and programming to date is limited. Where available, preliminary findings are reported. A research agenda is suggested.

VIEWER AVAILABILITY AND PROGRAM STRUCTURE

Gensch and Shaman (1980) suggest that viewer availability may be the best predictor of viewership. Using a trigonomic network share timeseries model, they were able to explain 85% of the variance in network viewing share across a year at 7:30 p.m., 50% of the variance at 9:30 p.m., and 44% of the variance at 10:00 p.m., using only an annual component which parallels changes in the amount of daylight. They propose that non-TV activities determine the size of total network viewing, and that behavioral assumptions about how viewers determine what to watch may require a subtle change. Rather than first selecting programs and then organizing other activities, individuals may first determine whether or not to watch television, and then what to watch.

Structural programming factors have consistently confounded attempts to predict network viewing behavior on the basis of content or program type preferences. Factor analytic attempts to isolate program types from viewing data find scheduling factors—day of the week, time of day, network channel, etc.—exerting more influence than program type on the resultant groupings of programs. Webster and Wakshlag argue that viewer availability to watch television "is the single factor most responsible for the absence of content-based patterns of viewing, as well as, the presence of structually defined viewing patters" (p. 438).

All the studies which have sought to define program types or predict actual viewership on the basis of preferences have thus far considered broadcast viewing situations, normally concentrating entirely on the networks. Thus, three different programs are available at any given point in time. It is the limited and variable structure of program options that makes predictions of network viewership so sensitive to structural factors. Cable television alters the available structure, expands program options, and provides more consistently available content types. In doing so, cable reduces the importance of viewer availability in terms of both time and content.

Cable's Program Option Structure

Cable channels tend to specialize (to different degrees) in some type of content, often making it available 24 hours a day. ESPN ensures that sports programming is always a viewing option; Reuters' Stock Channel permits constant exercise of the preference for watching stock news; Nickelodeon defines itself as children's programming. Unlike broadcast viewing, where those who prefer sports must wait until it is offered, much of the content structure is constant.

Channels available over cable can be aligned on a continuum of program predictability by content specificity. The networks, independent stations and superstations, and other general appeal channels would be the least predictable, offering different types of content, often different from day to day and time slot to time slot among days. Then there are generic content-typed channels, like ESPN or the movie channels, where a broadly defined program type is always available, although there is uncertainty whether the viewer will like the specific program (rugby versus baseball; *Clockwork Orange* versus *Mary Poppins*). MTV, offering short clips of rock music with accompanying video, and CNN, with a repeating newscast cycle, are more specialized. A viewer who likes news or likes rock music is fairly certain to find some preferred content quite quickly. Finally, there are channels so predictable, with repetitive, continuous content, that all they offer might be a clock or local weather.

Because of the greater variety and fixed structure of available content with cable, program choice should better reflect viewers' content preferences, particularly among those viewers who take advantage of cable's diversity and among those who have content preferences for which specialized channels exist. Cable may provide a better environment for assessing the impact of program type preferences.

Another structural impact of cable, which has more implications for *how* television is viewed than *what* is viewed, is program length. The networks traditionally offer hour and half-hour programs, and even the movies and sports events which exceed this tend to start and end on the hour or half-hour. On cable, some of the channels run at very different lengths. MTV provides short clips of 5–7 minutes. CNN cycles through a 45-minute news cycle. Text channels have cycles of varying length. Sports events and movies have variable end times. The weather channel is meant to be viewed for very short periods. This structural variation may alter the practice of watching in half-hour blocks. Viewers may join and leave programs at different points, in part on the basis of their own availability, having just completed viewing something else. Also, there is program content available during commercial breaks, when, with only the networks to choose from, there were only other commercials to turn to.

An analysis of interactive cable viewing data found that these cable channels are watched in different ways (Chapter 4). One-third of all channel changes to the networks resulted in a viewing stretch longer than 15 minutes. The average time per network stretch was 1 hour. MTV, on the other hand, resulted in stretch viewing from only one-tenth of all channel changes to that channel. MTV stretch viewing periods lasted about 50 minutes. CNN was in between: one-fifth of changes resulted in stretches, lasting an average of half an hour.

Similarly, an examination of *when* viewers change channels throughout a composite hour cycle (each of the 24 hours of the viewing day collapsed into a single hour) identified peaks of channel changing at the hour and half-hour. Slight increases were also observed near quarter-hour marks, conforming to expectations of the network program and commercial structure. However, channel changing was spread across the hour cycle, by no means limited to those predictable peaks. Rarely did changing for any minute in the hour drop below 70% of the rate expected by chance, if changes were evenly spread across all minutes of the hour.

Channel Loyalty

Channel loyalty is cited by Webster and Wakshlag (1983) as one of the routinely observed features of viewing behavior in broadcast studies, one which appears to predict a significant amount of the variance in viewership. They define channel loyalty as "the tendency of programs on the same channel to have a disproportionately large duplicated audience"(p. 434).

Channel loyalty, in models of broadcast network viewing, takes several forms. The "inheritance effect," also labeled "audience inertia," where viewers of one show on a channel are disproportionately likely to continue watching the next show on that channel, has been one of the strongest predictors. Inheritance effects have been found to occur primarily between adjacent programs, with effects ranging in size but accounting for less than 50% of the second show's total audience (Goodhart, Ehrenberg, & Collins, 1975). For programs on the same channel separated by more than one other program, effects are generally negligible.

Another way of defining channel loyalty has been the tendency for viewers to watch a particular channel, even when the program that used to be on it is replaced. In a study of the impact of network affiliation change by local stations, Wakshlag, Agostino, Terry, Driscoll, and Ramsey (1983) found that most viewers continue to watch news on the same *channel* even though the network news they were accustomed to watching had moved to a new channel.

In broadcast models, channel loyalty has been considered relatively content free (in part because the adjacent or replacement program may be totally different from the original show). Instead, channel loyalty is interpreted to be an outcome of stochastic, or "as if random" processes (Goodhart et al., 1975). An alternative or additional conceptualization of channel loyalty would be a "tendency to watch programs on one channel in preference to others" (Wakshlag et al., 1983, p. 53). With cable, there are three reasons to expect channel loyalty to be more pronounced. First, many of the cable channels specialize in particular program content. Thus, we would expect channel loyalty to cease to be a content-free function, and in fact to become confounded with "program-type loyalty." Second, with commercial television, programs have been traditionally considered a "free public good," costing the viewer nothing directly. With cable, special pay services (for the most part, movie channels) and tiers of special channels are available for an extra monthly fee. Viewers who are conscious of paying for certain channels may watch those channels more than they would were those channels available at no cost. Third, because of the large number of cable channels, identification of favorite or usual channels may be a means of simplifying the choice process, avoiding the need to evaluate all channels and instead concentrating on a subset of channels which are regularly considered. In fact, viewers have been found to watch a subset of channels regularly (Nielsen, 1983c), which has been labeled an individual's "channel repertoire" (Chapter 2).

Channel Repertoire

The impact of cable television (on broadcast television) is often reported in terms of differences in viewership to channel types (e.g., Webster, 1983). The channel types reported tend to be based on channel source, rather than channel content (e.g., network channels, independent, pay movie, PBS, and basic cable). Webster presents an analysis of viewing shares with and without cable across 24 markets. Weekly viewing shares among cable viewers are lower for local network affiliates and public TV stations. Pay cable channels attract a 14%–20% share in pay cable homes. Distant stations draw a 10% share in the largest market and as much as 46% in smaller markets. The remaining cable-only channels account for very little viewing time (less than 10%). These kinds of aggregate comparisons do not address the impact of cable on individual viewers. They demonstrate that networks still attract the plurality of viewing share, and that the individual basic cable channels do not attract appreciable aggregate viewing shares. To conclude that, therefore, cable does not alter viewing choice process or outcome may eventually be warranted; but studies of individual and household use of cable suggest that cable has substantial impacts. For example, although viewing *shares* (percentage of overall viewing time) of public television and independent and network stations are lower among cable households, the cumes (percentage of households that watch the channel) are higher for cable (Webster, 1983). Cable viewers watch more different channels, if for less time per channel.

Cable viewers may watch more different channels, but they watch far fewer channels than the total number available. In cable systems with 20 or more channels available, an average of about 10 different channels are watched regularly (Nielsen, 1983c; Chapter 4). Other studies report fewer significantly viewed channels, often in cable systems with lower channel capacities, but in each case, cable viewers do watch more different channels (Webster & Agostino, 1982; Webster, 1983; Television Audience Assessment, 1983).

Individual use of each channel in a 35-channel system was assessed in a study reported in Chapter 2. Of 35 channels carried, only the three local network affiliates were regularly watched by 50% or more of the cable subscribers surveyed. HBO, WTBS, and a local independent were watched by 40%–50%. Nine of the 22 other channels available only with cable (ESPN, MTV, CNN, Nashville, Cinemax, USA, CHN, Nickelodeon, and an "inspirational" channel) were watched by one-tenth to one-third of viewers. Thirteen cable channels were watched by less than one-tenth of subscribers. This pattern suggest that one individual's repertoire of 10 regular channels may be very different from another's repertoire.

Cable viewers have a limited set of channels that they watch regularly. The broadcast networks (or at least one or two of the networks) tend to be common across channel repertoires. But specialized cable channels are watched by highly fragmented audiences. About 10% of cable viewers surveyed had channel repertoires of two or fewer channels (Chapter 2). For 11%, the repertoire consisted of 12 or more channels. Viewers with small channel repertoires, and viewers whose repertoire includes only broadcast channels, are not likely to be taking full advantage of the content available on cable which would appeal to them. Conversely, a viewer whose repertoire consists of CNN, Reuter's Stock Channel, and AP News is using cable to specialize in highly preferred content, but possibly misses out on other preferred options.

We have suggested that cable's expanded and more consistent program-availability structure decreases the importance of viewer availability to watch television, and should increase the importance of channel loyalty and program-type preferences in predicting program choice. Next we turn to viewer awareness, and the actual choice process.

VIEWER AWARENESS AND PROGRAM CHOICE PROCESS

While the broader and more constant program structure of cable decreases the importance of viewer availability in predicting program choice, it increases the importance of viewer awareness. Webster and Wakshlag (1983) point out that both uses and gratifications models and the program type preference models of selective exposure assume perfect viewer awareness of program alternatives. Similarly, predictions that certain affective or arousal states will lead to exposure to certain content presuppose awareness or willingness to search for unknown program content. With cable, the assumption that, each time viewers select a program, they are aware of and weigh all program alternatives to select a most preferred option is untenable. There are simply too many options. Even on a general basis, cable subscribers are not very aware of the different channels available to them over cable, let alone the different programs.

Arbitron (1983) reports that "cable focus groups conducted in June, 1981 revealed that cable subscribers are not aware of all the services available to them, or even of what service they're watching at any given time" (p. 13). In door-to-door interviews, viewers were able to correctly identify an average of 9 of 35 available channels by channel number or location on the channel selector, and were aware of one additional service of uncertain location (Chapter 2). Twenty-three percent of respondents were able to identify only 0–3 channels. A common response from them was "I don't know, I just watch TV"; they had little knowledge of what local broadcast channels existed, let alone the specialty cable offerings. Another 23% correctly named 14–27 channels, and the remainder identified between 4 and 13. The number of channels a viewer could identify was highly correlated (.60) with the size of their channel repertoire. Respondents generally were familiar with more channels than they watched regularly. Chapter 2 described and tested a model of program choice, developing the concepts of orienting search, reevaluation, and viewing style.

It is important to note that not all television viewing invokes the proposed model of program choice. About half the time cable viewers know what they will watch before they approach the viewing situation, while, the other half of the time, programs are chosen at the time of viewing (Television, 1983). When viewers turn to television to see a specific, preselected program, the choice process specified here does not occur.

Orienting Search

There is evidence that some cable viewers attempt to become aware of program alternatives. Two studies reported greater guide use among cable households

(Chapter 13; Television, 1983), although one study found no significant differences (Sparkes, 1983b). Guide use with cable was bimodal, with one-third of viewers almost always checking a guide before watching television, and one-third almost never doing so (Chapter 2). Another means of becoming aware of program options is scanning the channels themselves. Cable viewers are more likely to scan. One-third of noncable viewers, 47% of basic cable, 49% of single-pay and 52% of mulitpay subscribers reported almost always or often scanning channels before deciding what to watch (Television, 1983). These efforts don't appear to be uniformly successful. Despite guide checking and channel scanning, two-thirds of cable viewers were aware of few or no program alternatives at the time of viewing, while less than one-third were aware of what was on "all" or "most" other channels (Television, 1983).

Examination of the different methods of channel scanning suggest different impacts on awareness and channel repertoire, depending on the method used. Three dimensions of an orienting search have already been identified: *processing mode* (automatic versus controlled), *search repertoire* (elaborated versus restricted set of channels checked), and *evaluation orientation* (exhaustive versus terminating searches) (Chapter 2). These dimensions appear to reflect consistent viewer approaches to the choice process.

Thirty-six percent of respondents reported scanning 26 to 35 channels in order (with a mode of 35, encompassing 20% of the respondents). These were classified as using the automatic processing mode. Fifty-three percent checked between 0 and 5 channels in order, suggesting a controlled search of specific channels in an order of their own design.

An orienting search can be elaborated, including most or all of the channels available on the system, or it can be restricted to a small number of options. Elaborated searches have better potential for viewer awareness, and require more information processing effort. Forty-two percent of cable viewers reported scanning 34 to 35 of 35 channels possible. Fifty-five percent checked 0 to 14 channels ($\bar{x} = 5.9$), evidencing restricted searches.

A *terminating search* is defined as a search in which the viewer abandons the search when the first option that meets some minimal standard is located (in economic terms, a "satisficing" approach). An exhaustive search includes scanning of *all channels which comprise the individual's search repertoire*, followed by a return to the *best* alternative. Half the cable subscribers surveyed normally engaged in terminating searches, 46% used exhaustive, and 4% weren't sure (Chapter 2).

Viewers were found to have consistent, habitual orienting search patterns. After describing their pattern, more than half of the respondents claimed that they "almost always" used that same approach. Three-fourths of them used the pattern they described at least half of the time.

Controlled searchers tended to check a restricted number of channels ($r = .97$) and more frequently used terminating searches ($r = .23$). Terminating searchers used their search pattern significantly less of the time (66%) than exhaustive searchers (77%). Automatic searchers checked an elaborated set of channels and were most likely to use exhaustive searchers (r = .25). To summarize, the orienting search dimensions are highly related (Chapter 2). The search pattern likely to elicit the most viewer awareness would be the automatic, exhaustive, elaborated search. Viewers who are aware of all program options should be more likely to find one which best suits their need if it exists on the cable system, and should, therefore, most clearly evidence more correct predictions of selective exposure research and program preference models.

Re-Evaluation

One definition of an *active viewer* is a person who engages in demanding orienting searches to select programs. Once a program has been selected, that critical active processing may cease until the chosen program ends, or frequent re-evaluation of the chosen program can continue with the intent of changing to other alternatives if it is not up to expectations. Again, viewers who actively engage in re-evaluation are likely to watch programs more consistent with their program preferences or need states.

Cable viewers change channels more often at commercials between and during programs than noncable viewers (Chapter 13; Television, 1983; Chapter 2). For both cable and noncable, channel changing occurs more often during shows which are watched once or occasionally than during viewers' regular shows (Chapter 2).

Cable viewers have been classified into four levels of channel-changing activity by Guttman scalogram analysis, each group comprising about one-fourth of the viewers (Chapter 2). The least active group rarely changes channels; group 2 does so primarily between programs; group 3 also changes when commercials come on, during a show; and the most active group changes in the middle of a show, other than at commercials as well. All of the channel changing measures were significantly related to each other. The passive viewers average perhaps one channel change every 90 minutes during regular shows and one change every 45 minutes during nonregular shows. They search for something to watch less than twice a day, and only 9% of their members ever watch more than one show at a time. The most active reevaluators change channels three out of every four commercial breaks, change two of every three periods between programs, and change in the middle of 44% of the shows they do watch. They report changing channels 10 times an hour during regular shows and 12 times during nonregular shows, and two-thirds of them sometimes watch more than one show at a time.

Viewing Style

Half the cable subscribers surveyed generally do not even consider changing channels until the program they originally selected has ended. They approach the viewing situation from what might be termed a "program orientation." The other 50% use commercial breaks as an opportunity to examine other program options, and half of them (24% of subscribers) will even change in the midst of program content, exhibiting the weakest program orientation and greatest activity level while viewing.

A viewing style analysis of interactive household viewing data tends to support these systematic differences in viewing style and program orientation. Three modes of viewing were identified and operationalized: channel sampling or scanning (watching a channel or a "string" of channels each for 4 minutes or less); extended sampling (watching channels 4–15 minutes—less than a full program, but more than a brief scan), and program or "stretch" viewing (conservatively defined as viewing a channel 15 minutes or longer, to avoid underestimating program viewership). On the average, scanning comprised 8% of total viewing time, extended sampling 10% and program viewing 82%. Scanning and extended sampling were consistent across days (average correlations = .70), were highly correlated with each other ($r = .92$) and exhibited strong individual differences across households.

The choice-process variables described and proposed here have been found to be highly interrelated, in ways consistent with information-processing expectations. There are viewers who engage in elaborated, exhaustive, automatic orienting searches. They also tend to be the ones engaging in more frequent reevaluation. They are familiar with more different channels on the cable system. They have larger channel repertoires, and they watch more cable channels. They tend to be younger adults, consistent with the information processing literature which identifies young adults as possessing the best information processing skills (Watt & Krull, 1974). These viewers are the most aware of programming options, and presumably should exhibit the best fit between preferences and needs and program exposure.

The information-processing model suggests that viewers with different information-processing approaches will watch cable in different ways and this is supported by the consistent interrelationships. The wide individual differences further suggest that an individual, information-processing variable may differentiate viewing styles.

VIEWING GROUP

Webster and Wakshlag (1983) point out that most models of program choice ignore the role of group viewership, despite the fact that most viewing occurs in groups, and group composition affects the program selected. They detail the possible influences of group viewership on the choice process in general.

> First, an individual's specific program preferences may have to operate through the viewing group to determine a program selection. Second, a program choice, perhaps the result of some intense viewer preference, may affect the group composition. Finally, the nature of the group itself, may be a cause of specific program

preference. For instance, a parent wishing to view with children, might prefer a program that is suitable for the group, but is otherwise unappealing. (p. 441)

But what impact does cable have on group viewing? The choice process is much more complex with cable—as discussed throughout—because of so many different choices. Because some of the channels appeal to particular and specific content preferences (e.g., sports, news, Nickelodeon, Playboy Channel), more conflict over what to watch might be expected. There is a greater likelihood that each person in a viewing group would select a different optimal program, were they aware of all viewing alternatives, and were they to select the best one for their own needs. Somewhat surprisingly, one study which compared cable and noncable group decisions found no significant differences in how often people in their households agreed on what to watch, the infrequency with which they watched something other than what they would have watched alone, and the in- frequency with which they left the viewing situation rather than be exposed to disliked programming (Chapter 13).

Cable viewing with a remote channel selector might be expected to increase the dominance of one individual (whoever holds the channel selector). Contrary to this expectation, there was no difference between cable and noncable channel selector dominance by the randomly selected respondents (Chapter 13). Within cable homes, females were less likely to report that they controlled the channel selector, and more likely to report that someone else changed channels when they wished they wouldn't (Chapter 2).

Cable viewing in a group might also be expected to evidence less channel changing than viewing alone. When viewing alone, there are few constraints on changing channels as the viewer becomes bored, or at commercials, or in the middle of a show. When viewing in a group, there is the distinct possibility that others might object. Again, contrary to expectations, when viewers were asked separately about channel-changing behaviors when alone and in a group, more channel changing was reported in the group situation than alone (Chapter 13).

In fact, viewers in cable and noncable homes both reported on the average more guide checking while watching, more channel changing between shows, and more channel changing at commercials in a group than while watching alone. One explanation is that there is more diversity in the group's viewing pref- erences, and the changes are to accommodate that. Another explanation is that individual differences in channel changing exist, but that, in a group, there is a tendency for the channel changer to dominate, with more resultant changes on the average in groups than across randomly selected individuals reporting on their own behavior when watching alone.

ACCESS TO CHANNELS

Cable also introduces means of accessing its multiple channels which have dif- ferent implications for influencing the program-choice process. These channel-access factors do not fit well into any of Webster and Wakshlag's (1983) existing categories, and should probably be added to their model.

Cable television typically provides viewers with a channel selector which is separate and different from the traditional built in TV set dials. Cable channel selectors range from small, hand-held digital units, to bulky push-button units, to combinations of the built-in TV set dial for Channels 2 to 13 and a separate unit for additional channels, and so on. The subscriber may instead substitute some other commercially available unit (e.g., remote infra-red selector, or a "cable ready" TV set). Different attributes of these channel selectors may affect viewers' channel selection patterns.

Remote Control

The most obvious factor is whether or not the channel selector can be comfortably accessed from some normal viewing position. Selectors with remote extension cords (or wireless selectors) greatly reduce the effort involved in changing channels. Increased ease of channel changing should presumably result in more channel changing with other elements held constant. Specifically, we would expect viewers with remote channel selectors to be more likely to engage in orienting searches which are exhaustive, extended, and automatic. They should be familiar with more channels, watch more different channels, and engage in more frequent re-evaluation, zapping commercials, changing in the middle of a show if it doesn't suit them, and watching more than one show at a time. In short, they should be more effective at locating the content available on cable that best fits their needs.

In 1982, the FCC commissioned a study of the impact of broadcast channel selector types on UHF viewership in noncable households. The comparisons included continuous versus discrete single and double built-in dials, as well as remote selectors. No significant difference was found in the number of quarter hours of UHF programming viewed across the different selectors. The study was based on diary data reported in 15-minute blocks, which may be too gross a measure to tap the increased channel sampling of UHF channels (and other channels) that the literature reviewed here would predict (Brenner & Levy, 1982).

A more recent comparison of remote and nonremote channel selectors examined orienting searches by cable and noncable viewers. No difference was found in noncable viewers: one-third with, and one-third without remote selectors almost always or often scanned channels to decide what to watch. A difference in the predicted direction was found among cable viewers, where 54% with remote selectors and 47% without remote selectors scanned channels, although the difference was less pronounced than the researchers expected (Television, 1983).

Two other studies considered the impact of remote selectors among cable viewers. One of them characterizes cable viewers with remote selectors as:

- watching less afternoon TV, prime time TV, and TV overall
- less likely to say they watch too much TV

- more likely to zap commercials during and between shows
- more likely to change channels during a show
- more likely to believe it's hard to decide what to watch
- more likely to report that others change channels when they wish they wouldn't (Chapter 13).

Watching less TV is consistent with the expectation of more selective viewing, assuming that selective viewers may turn the set off when nothing desired is on. Increased channel changing implies more frequent re-evaluation. Difficulty deciding what to watch could be consistent with an extended, exhaustive orienting search where the viewer seeks maximal information before selecting one alternative.

The second study specifically assessed the relationship of remote selectors with elements of the model proposed here. A remote selector was associated with significantly more re-evaluation, a more extensive orienting search, and a larger channel repertoire, in bivariate comparisons, as predicted. However, contrary to hypotheses, no significant relationship was found between presence of a remote selector and channel familiarity and exhaustive versus terminating searches (Chapter 2).

None of these studies were specifically designed to assess the impact of remote selectors. A universal problem across them is a self selection bias: noncable households and cable households in at least two of the sample cable systems had remote selectors because they were willing to pay for them.

Type of Selector

More subtle and complex than the question of remote control is the effect of different types of channel selectors. To exemplify potential impacts, some of the major selector types will be described and their implications for the choice process explored.

Older cable systems with limited channel capacity often still use the discrete dial built into a TV set for accessing channels. With a dial, channel order for orienting searches is essentially fixed. The viewer can choose whether to cycle clockwise or counterclockwise, but is forced to at least pass over every channel, in order, when changing. To the extent orienting searches are engaged in, they are likely to be elaborated (checking many channels) and automatic (following the numeric channel order). Given forced exposure to all channels in the rotation, channel familiarity might be expected to be greater. For reevaluation, however, changing channels at commercials or trying to watch more than one show at a time, with the specific intent of returning to the original channel shortly, can be quite cumbersome. Reevaluation might be expected to be reduced. Overall, the cumbersomeness of moving purposively from one channel to another, nonadjacent channel might inhibit selection processes and, therefore, viewer effectiveness at identifying content on cable which best suits their needs.

When older 12-channel cable systems add new channels, they frequently provide a unit separate from the television dial to access new channels, but continue to use the set dial for the original channels. The necessity of moving between two modes of access for different groups of channels may encourage viewers to concentrate on one of the two groups, limiting channel familiarity, size of orienting search, and channel repertoire.

A common cable channel selector is the push-button, normally consisting of one row of 12 buttons, and a three-position switch on the side, making available 36 channel positions. On push-button units, it is easiest to change channels either horizontally, within a row, or vertically, between rows. This may influence orienting-search order. Where a channel is positioned and the popularity of vertically and horizontally adjacent channels may influence channel use. With a push-button unit, channels are accessed spatially rather than numerically. Pretesting of an instrument to assess channel familiarity using a list of channel numbers found the more industrious push-button selector respondents using their fingers to try to match their spatial memory of where the channel was with the number to which it should correspond (Chapter 2). There may be a tendency for viewers to specialize in one row. Channel changing is easy, from one specific channel to another, or straight down a row or column. Channel familiarity as well as channel repertoire might be expected to be larger. Also, viewers with a strong spatial orientation may be more skillful and more frequent channel changers.

Digital channel selectors, on which channels are accessed by entering numbers of a keypad, likely encourage a different approach. Here, an orienting search is likely to occur either from known channel number to known channel number (controlled) or in numerical order (automatic). An extended automatic search in a 35-channel cable system would require entering 70 digits (two digits per channel). Considerable more effort is required than with a push-button unit to scan all channels, suggesting a restricted search, less channel familiarity (but knowledge of channels, by channel number, of those that are known), and perhaps a smaller channel repertoire with digital selectors. Re-evaluation should be equivalent or enhanced, due to the ease of moving from specific channel to specific channel. Viewers comfortable with numbers and calculators should change more often and be more skillful at locating desired content.

Many of the digital selectors include a built in function for "preprogramming" 10 channels, such that the 10 programmed channels may be automatically scanned, in the programmed order, at the press of a single button. Pretesting results suggest that few cable viewers with these devices are even aware of the programmable function. Those that are aware and do use it are likely to have a very fixed orienting-search pattern limited to 10 channels, and always following the same order.

A comparison of digital remote selectors and short-cord push-button units found equivalent channel changing for the purpose of turning to specific shows and to see what else was on. Digital viewers changed more often when bored, to avoid commercial, for variety, and to watch more than one show (Chapter 2). These results are confounded with the impact of remote control. No systematic examination of the impact of cable channel selector type has yet been undertaken.

CABLE AND PROGRAM SATISFACTION

Most of the premises set forth in this chapter have been based on the assumption that cable, by making available more and different programming, has the potential to enhance the correlation between viewer preferences and needs, and program exposure. Yet most of the research which has been conducted on cable has concentrated on viewership of channels and channel types rather than programs and program types. A few studies have examined aggregated program-type viewership. Cable subscribers, on the average, perceive themselves as watching more movies, sports, weather, national news, and news in general, and less local news, fewer talk shows, and fewer soaps since they subscribed to cable (Chapter 2). The reported change in program-type viewership suggests that there may be some difference in program exposure due to cable.

A study of diary-reported viewing by cable and noncable respondents found fewer, and somewhat contradictory, differences. Television Audience Assessment (1983) measured program type exposure over a 2-week period, finding that pay cable subscribers watched more movies and sports and less public affairs and news shows than basic cable or noncable. All cable viewers watched more sports. The following program types were not different by cable subscription: comedies, action adventure shows, dramas, variety-entertainment, and miscellaneous. The availability of cable in homes altered viewing of only a few program types.

News viewership has been studied in more detail. Henke, Donohue, Cook, and Cheung (1983) report less-frequent news viewing among cable subscribers, and particularly less network news. In Chapter 16 we find that, among cable viewers, 25% of their news viewing was on cable-only channels. Of the 20 minutes spent with cable news channels or programs, one-fourth occurred in periods when no broadcast news was available (demonstrating greater exercise of program type preference with cable). The remainder of news viewing occurred when broadcast news was also available (greater exercise of specific program preference, within an already available program type). Similarly, Hill and Dyer (1981) report that 39% of cable subscribers watched local news on a broadcast channel, while an additional 16% watched local news on a distant broadcast channel available to them only with cable.

Implicit in the premise that cable offers, and viewers avail themselves, of more programs which satisfy their needs is the corollary that cable viewers will be more satisfied with programs they do choose to watch. Television Audience

Assessment (1983) used their qualitative audience appeal and audience impact scores to compare cable and noncable viewer ratings of the programs they watched over a 2-week period. Both groups rated news programs, entertainment shows, and movies they watched at about the same level. Noncable viewers were neither more nor less satisfied with broadcast programs than cable viewers. Cable respondents did rate the cable programs they watched at the same appeal level as the broadcast programs they watched. Overall, satisfaction with television was significantly higher among cable respondents, but program satisfaction was not different. The researchers conclude that it is the individual program, not the means of delivery, that determines appeal and impact ratings.

The program appeal index measures "overall entertainment value" and provides an estimate of whether viewers will *plan ahead to watch a show again*. Program impact measures intellectual and emotional stimulation, involvement, and likelihood of viewers *watching commercials placed within a program*. Thus, these two satisfaction indexes are confounded by viewing style behaviors which differ between cable and noncable viewers. Studies which more clearly define and measure program satisfaction and viewer-need satisfaction are needed before definitively concluding that cable does not result in greater viewer-need satisfaction. Program type loyalty and other need-based program choice predictions should be assessed on an individual level, under more controlled comparisons, isolating the impacts of number of program options and consistent content structure.

DISCUSSION

Cable television differs from traditional broadcast television along a number of fundamental dimensions which have implications for program choice. Some of the differences have the potential to improve predictions of selective exposure. Others further confound the issues. The most obvious change is that the number of channels increases, often dramatically, from a range of one to 10 channels available over the air, to as many as 108 or more channels over a cable system. Second, the structure of programming available over the three broadcast networks is highly variable by program type (movies, sports, talk shows, comedies, action adventures, etc., all at different times, but not all the time), while the time structure for broadcast programming tends to be fixed, with programs typically lasting 30, 60, 90, or 120 minutes. On cable, the situation is reversed. There are many channels specializing in a particular program type (e.g., country music or news) such that much of the programming content is fixed. Program length is highly variable, with some "programs" lasting a few minutes and others for hours. Third, access to cable is different. Cable offers remote extension cords, and channel selectors other than the traditional dial. Finally, television programs have been considered a "free" public good. With cable, some of the channels are available for a group price; others are charged for individually.

This chapter has begun to explore the implications of these cable-broadcasting differences for program choice. Many of the propositions suggested here have yet to be tested, or are based on limited research. The research agenda is extensive.

We have suggested that the expanded program options and more consistent content structure with cable will decrease the impact of viewer availability to watch television, and of scheduling factors, with a result of improving program choice predictions based on program type preferences. Suggested also is the possibility of enhanced predictions for all need-based models of program choice. These expectations must be tested empirically.

Cable channels have been said to have varying degrees of content specialization and program length cycles, and to be watched in different ways, by varying numbers of cable subscribers. Channel loyalty may be an even stronger predictor of viewership with cable, perhaps confounded with program type loyalty. How much of the variance in viewership does cable channel loyalty explain? Can channel use profiles be created? Are there channel types, distinguishable on the basis of use? What factors (structural, content, placement) determine channel use?

Channel repertoire has been defined as the set of channels which a viewer regularly watches. What causes a large or small channel repertoire? How does a channel repertoire develop? How quickly? How stable is it? To what extent does a channel repertoire limit viewers from watching (or even knowing about) content on other channels? There has been some evidence that individuals in a household share the same or similar channel repertoires. To what extent? How well does an individual's channel repertoire reflect his or her individual preferences? How can channel repertoire be used to define audience fragmentation? How can channel repertoires be characterized to describe a viewer's television fare?

Viewer awareness has been proposed as a key link between viewer needs and selection of the best alternative program. This can be tested empirically by manipulating awareness. How are guides used? Can a more effective guide be developed which will alter awareness and program choice? What impact would educating cable subscribers about the channels on their system have on viewing? Will increased awareness alter viewing patterns? Does channel selector type or remote capability influence awareness?

A model of program choice process has been proposed which posits adoption of an indefinite goal, capable of being satisfied in different ways and to different extents by more than one program option, where selection of a program option is based on information generated within the problem-solving process about what alternatives are available and which needs they best serve. Only one study has tested this model, and it defined and linked general channel familiarity, orienting search pattern, re-evaluation tendencies, and channel repertoire size (Chapter 2). Laboratory study is needed to assess whether goal clarification actually occurs during an orienting search. Do viewers evaluate alternatives with an

"ACCEPT/REJECT/CONSIDER AND CHECK OTHERS" framework? An inverted U-shaped relationship between uncertainty and the attractiveness of a situation is consistently found in research on information processing abilities and preferences (e.g., Munsinger & Kesson, 1963; Eckblad, 1966). Does this same distribution occur for channel-selection processes? If so, the expectation that program choice is a function of information-processing abilities may be supported.

There appear to be systematic relationships among the various choice-process measures. An orienting search which uses exhaustive evaluation is also likely to include an elaborated search repertoire and an automatic-processing approach. That composite behavior pattern is associated with ongoing reevaluation. The interrelationships need to be better defined. A small set of viewer types, differentiated by viewing style or program-choice style may exist, and that classification can be used in further research on selective exposure. Orienting search has been measured only by self-report thus far. How valid are those data? What is the relationship between choice process and outcome? In terms of channel repertoire? Viewer satisfaction?

The choice process model is based on one individual selecting a program. What happens in the group viewing situation? There has been evidence of more channel changing in a group than alone. Can the model be applied to group viewing? Do goal-clarification and orienting searches occur in that setting? If so, it may be a good opportunity to study the process, where it is more overt, perhaps even verbally discussed. If not, how then is the group choice made? Do children engage in the same choice process as their parents? Do their viewing styles develop?

Many of the propositions advanced in this chapter can be linked to other bodies of research. For example, research on the need for arousal (e.g., Berlyne, 1960; Zuckerman, 1979; Donohue, Palmgren, & Duncan, 1980) may provide an explanation for differences in program-choice styles. Those with high needs for arousal may be the most active viewers. Cortical as well as limbic arousal research should be considered. Studies which have examined attention to the screen (e.g., Anderson, Alwitt, Lorch, & Levin, 1972; Anderson & Levin, 1976; Husson, 1982), in situations where viewers did not have the opportunity to change channels, may be extended to cable channel changing situations, to determine how closely attention and channel changing are motivated by parallel-program or attention attributes. Research on selective exposure, information processing, and decision making are also primary candidates for integrative efforts. What has been presented here is a beginning. Perhaps it merits further exploration.

## CHAPTER 4

Cableviewing Behaviors:
An Electronic Assessment

Carrie Heeter, David D'Alessio,
Bradley S. Greenberg, and D. Stevens McVoy

Generally, cableviewing is presumed to be a more active than passive behavior. Evidence supporting that description has been limited to self-report measures. This chapter examines actual viewing and choice processes as they occur in cabled homes, using minute-by-minute interactive cable viewing data on: how often and when viewers change channels; how different kinds of channels are viewed; and what constitutes active viewing in terms of channel changing.

Research attention has begun to focus on differences in how people with and without cable view TV from the standpoint both of being able to predict what they will watch, and of understanding the levels of selectivity and active viewing that occur with cable. A well-documented difference is increased channel sampling and channel changing with cable. Sampling or circulation of specific channels (watching that channel for 5 minutes or more in a week) tends to be higher among cable households. Although public television shares (percentage of viewing time) in cable homes are smaller than in broadcast homes, the cumes (percentage of households that watch the channel) are higher (Agostino, 1980; Nielsen, 1983c). Similarly, where significant differences have been found between cable and noncable household cumes for network affiliates and independent stations, circulation was higher among cable households (Webster, 1983).

Cable households watch more channels, although how many more varies across studies. Nielsen reports that about 10 different channels are regularly watched in cable systems with 30 or more channels (Nielsen, 1983c). Other studies report lower numbers of significantly viewed stations, but are not or may not be limited to 30 + -channel systems (Webster & Agostino, 1982; Webster, 1983; Television, 1983). In all comparative studies, the number of channels viewed in cable households was significantly higher than in broadcast households. How viewers cope with the diversity available from cable is an interesting question. Arbitron reports that "cable focus groups conducted in June, 1981 revealed that cable subscribers are not aware of all the services available to them, or even of what service they're viewing at any given time" (Arbitron Ratings, 1983, p. 13). In door-to-door interviews in a 35-channel system, cable respondents were able to correctly identify an average of nine different channels by location on the channel selector or channel number (Chapter 2).

Cable viewers engage in more random searching of channels and are more apt to "zap" commercials (Chapter 7). A two-system study found that 17% of cable viewers reported almost always changing channels during programs, and 39% almost always zapped commercials, compared to 6% and 13% of noncable viewers (Television, 1983). Cable subscribers with remote channel selectors were significantly more likely than cable subscribers without remote control to report that they zap commercials between and during shows, change channels in the middle of a show, and believe that it is hard to decide what to watch (Chapter 7).

All these data that purport to describe aspects of cable viewing styles and processes originate from self-reports, with no independent, external validation. Validation of many of these behaviors is possible by electronic measurement in interactive, two-way cable systems. This study supplies a portion of that validation effort: It applies behavioral definitions to several cable viewing variables. In doing that, we propose to outline parameters of cable viewing styles for subsequent inclusion in less-descriptive and more-theoretic examinations of cable viewing. For example, it is anticipated that, by identifying what kinds of channel sampling, scanning, full program viewing, and intermediate ranges of viewing occur, and the consistency of these behaviors across time, individual or household viewing styles can be defined. Subsequent research can then assess who behaves this way and why. Next, viewing behaviors for different types of cable channels are examined. This identifies the extent to which overall styles vary for different kinds of cable services. Finally, the issue of channel changing is addressed precisely in terms of when and how often it occurs across a day.

Thus, this study develops and operationally defines concepts from viewing data, heretofore explored solely from verbal reports. Those concepts in turn are expected to generate more heuristic examinations of the cable viewing process.

METHODS

Coaxial Communications, owner and operator of the cable system franchised to a Florida community, wired a random sample of 197 subscriber households for interactive audience measurement experiments. For one week in June, 1982 (Wednesday, 6/16, through Tuesday, 6/22), the headend computer in the cable system scanned channel changing behavior in all 197 households every minute, continuously for 1 week, storing the *time of each of the change*, the *channel changed to* and the *household identification number* whenever a change in channel occurred or the set was turned on or off. The data were transferred to computer disks, and then to magnetic tape with cooperation from Coaxial Communications, and shipped to the Communication Technology Laboratory at Michigan State University.

The Florida cable system carried 34 channels, including two pay services (HBO and the Movie Channel), four network stations, four independents, three superstations, two PBS stations, four local access channels, and 15 other cable-only channels. About 58% of the sample subscribed to both pay movie channels, 16% to neither, and the remainder were divided between the two. All households were equipped with remote, hand-held, digital channel selectors. To change channels, the desired channel number must be entered on the numeric keypad.

At any minute, a household can be tuned to one of the 34 channels, or the set can be off. Thus, for each day, there are 35 channel-state variables $\times$ 1440 minutes in a day $\times$ 197 households, or about 10^7 data points per day.

In order to extract cableviewing patterns, methods of data reduction must be developed, which depend on the questions being asked. For example, viewing can be *aggregated across households*, summing and comparing viewing time, number of channel changes, number of different channels watched, etc. Viewing can be *aggregated across channels*, to define channel profiles in terms of amount or time viewed, number of different times selected, number of different households viewing the channel, etc. Viewing can be described by *daypart*—the number or percentage of households watching a specific channel or changing channels at different periods in a day. This chapter reports data reduction approaches that describe and define the process of watching cable television. Computer programs were written to calculate the viewing measures.

Development of Viewing Measures

The traditional focus of audience measurement has been to predict or describe *how much* a given channel, program, or program type is viewed. The interactive cable data permit examination of *the ways* in which channels or shows are viewed, providing more detailed information than sample aggregate time shares. Three modes of watching were anticipated to occur: *sampling or scanning*, where a channel is watched for a very brief period in which the viewer rapidly examines the content; *program viewing*, where one or more shows are viewed in their entirety; and *extended sampling*—a cross between sampling and full program viewership, where a more careful, lengthy evaluation of content occurs, short of watching entire shows.

We graphed and visually examined a subset of 30 households' viewing for several presample data collection tests. In some households, fairly extended periods of channel sampling occurred, lasting as long as 10 to 15 minutes, during which a *string of channel changes* was made, but no channel was watched more than a few minutes at a time. Rather than occurring in isolation, in brief periods of changing between shows, or at a commercial, these extended strings of changes appeared to be fairly common.

A computer program was written to cumulate the number of recorded blocks of viewing time across all households by duration (e.g., 400 viewing blocks of 3 minutes occurred, 100 of 4 minutes, and less than 100 of longer time periods) for each of the sample days. This output was used to assist in operationalization decisions for the three modes of viewing expected to occur.

There was a sharp drop between blocks of 1 and 2 minutes, and a leveling off after 4 minutes. Viewing periods of 1 to 4 minutes on a given channel were selected to operationalize scanning, or *string* viewership.

The average amount of time channels were watched consecutively in the first viewing day examined was 15 minutes. This corresponds to the smallest time block for which commercial ratings services typically report viewing. Channel viewing in blocks of 5–14 minutes was the operationalization used to represent the intermediate level of viewing commitment—less than a full program, but more than cursory scanning. These were labeled *mini-stretches*. Without an elaborate computer program identifying schedules for all shows on each channel with specified lengths, it is impossible to know precisely from our data whether a household watched an entire show. The amount of time a channel is viewed permits an estimate of that likelihood. To consider viewing blocks of 15 minutes or more as an indication that an entire show was viewed errs in the direction of overestimating full show viewership. In this first attempt to utilize interactive data to examine the extensiveness of channel changing with cable, it was deemed more important to conservatively define string and ministretch viewing. Viewing blocks lasting 15 minutes or more were labeled *stretches*. Frequency of occurrence of these viewing measures will be presented in the results.

Error Rate

Errors during data collection could occur at the individual household's interactive channel selector (affecting only that household), at the cable headend computer (affecting one or all households), or at the "code operated switch" which regulates the flow of two-way information for an entire section or "block" of households (affecting the entire block of households). Errors were recorded in the same manner as channel changes by the headend computer. When an error condition was first detected, the time and error condition code were stored with the household ID. Subsequent monitoring searched for a change in status from error to readable channel or set-off condition. The period between first identification of an error condition and correction of it is thus known. Among three sample weekdays, 172 households which had consistently low error rates (.2%–1% of viewing time) across days were selected for this analysis.

RESULTS

The analyses identify household, channel, and daypart viewing profiles for cable.

Household Analysis

Household string, ministretch, stretch, and overall viewing were examined by viewing time, period of viewing per channel change, and the number of different channels selected.

Overall viewing. The typical household, averaged across three weekdays, had the TV set on for 7 hours and 44 minutes. During that time, they changed channels 34 times (4.4 changes per hour). Within those 34 channel changes, 10 different channels were selected at least once for 1 minute or more.

Table 1 contains average household viewing statistics by sample day. Total amount of viewing ranged from 7 to 8.5 hours per day, with 5%–7% of households not watching at all. The average household correlation between daily viewing times across days was .71. Number of different channels selected was very consistent, ranging from 9.5 to 10, with an average correlation of .66 across days. The consistency provides some support for the concept of a "household-channel repertoire," or subset of channels from which a household typically selects its viewing fare.

String viewing. String viewing comprised 8% of all viewing time and 64% of all channel changes. All but 13% of the households (half of which did not watch television) watched at least one channel for 4 minutes or less each day. The average amount of time spent per day in strings of changes was 35 minutes (range = 34–37 min). The maximum observed time spent viewing in strings by a single household ranged from an extreme of 4 hours 39 minutes on Wednesday to 7 hours and 21 minutes on Thursday. Examination of these data found few or no error codes, inclusion of numerous channels, and variable periods between changes, suggesting purposive changing rather than system error. Amount of string-viewing time was very consistent, with an average correlation of .71 across days.

String viewing included an average of 21 to 23 channel changes per day ($\bar{r}$ = .70) and the maximum ranged from 172 to 308 changes. These changes occurred in 6.6–7.5 separate periods of frequent channel changing ($\bar{r}$ = .74), lasting an average of 5 minutes and including three channel changes per string. Seven to eight *different* channels were selected in strings (fewer than the overall 10 channels watched in a day). The distribution of string viewing periods identifies string viewing as a persistent, consistent, and fairly common cable-viewing behavior.

Ministretch viewing. Ministretch viewing comprised 9%–11% of all viewing time and 16% of all channel changes. Forty-four minutes per day were spent watching channels for 5 to 14 minutes at a time, within which five channel changes (among three different channels) occurred, for an average of 8.2 minutes each. The maximum range for viewing in ministretches by a single household was 4:55 to 4:59, and ministretch viewing time and number of occurrences had average correlations between days of .69.

Stretch viewing. Amount of stretch viewing showed the most variation across days, ranging from 5:48 to 6:54. Stretch viewing consistently accounted

Table 1.   Household Viewing

	Wednesday Mean	Thursday Mean	Tuesday Mean	Overall Mean
Overall Viewing				
Total Time	7:06	8:26	7:40	7:44
# Channel Changes	32.4	35.6	32.8	33.6
# Different Channels	9.5	10.0	9.5	9.7
String Viewing				
String Viewing Time	:34	:37	:35	:35
% String Viewing	8.0%	7.7%	7.9%	7.9%
# String Channel Changes	20.8	22.8	21.0	21.5
Time/String Change	:01:36	:01:36	:01:42	:01:38
# String Periods	6.6	7.5	6.8	7.0
Time/String Period	5.1	5.0	5.2	5.1
Changes/String Period	3.2	3.0	3.1	3.1
# Different String Channels	7.5	7.9	7.3	7.6
Ministretch Viewing				
Ministretch Viewing Time	:44	:44	:45	:44
% Ministretch Viewing	10.5%	9.2%	10.0%	9.9%
# Ministretch Channel Changes	5.4	5.4	5.3	5.4
Time/Ministretch	:08	:08	:09	:08
# Different Ministretch Channels	3.0	3.1	3.0	3.0
Stretch Viewing				
Stretch Viewing Time	5:48	6:54	6:20	6:21
% Stretch Viewing	81.3%	83.0%	82.0%	82.0%
# Stretch Changes	6.2	7.4	6.5	6.7
Time/Stretch	:56	:56	:59	:57
# Different Stretch Channels	4.0	4.5	4.2	4.2

Means are averaged across 172 households. Overall mean column reports the mean of means across days.

for 81%–83% of viewing time and 20% of channel changes. Households averaged 6.7 separate stretches of viewing a single channel for 15 minutes or longer, each lasting an average of 57 minutes, lending some credence to our assumption that most stretches consisted of viewing entire shows. Four different channels were watched in stretches per day. Stretch viewing time was correlated .70 across days, and number of stretches correlated .69. The maximum number of stretches ranged from 18 to 20 per day.

Channel Analysis

Viewing behavior in terms of channel types was analyzed next. Six major channel types were defined: *network* (four channels affiliated with one of the major broadcast networks), *pay movie* (HBO and The Movie Channel), *superstations* (WGN, WOR, and WTBS), *area independents* (four channels), *general satellite* (six satellite programmed channels targeted for a broad cable audience, with some content specialization, e.g., MTV, CNN, ESPN), and *other* (a miscellaneous category grouping the remaining 15 cable channels). Some analyses include a breakdown of the miscellaneous channels, including *automated* (five character-generated channels with repeating text, e.g., 24-hour weather and shopper's guide), *PBS* (two area public television stations), *specialized cable* (four cable programming services targeted to a small, specialized audience, e.g., the Christian Broadcasting Network and Spanish International Network), and *access channels* (local origination, public, government, and educational access).

Table 2 presents household viewing data, aggregated across days, by channel type. The number of *different* channels from within that channel type that the average household watched in a day is presented. Proportionately, by channel type, the typical household watched two to three network stations, one pay movie channel, three superstations or independents, two general satellite channels, and an additional one to two channels from the remaining 15 less-popular options. The number of changes per household describes the allocation of channel changes by channel type, for an "average household," per day. Similarly, viewing time is the amount and proportion of time the average household devoted to each channel type.

The four network stations on the cable system received the largest share of channel changes and viewing time. Households checked or watched a network station 10 times a day, for 3 hours and 22 minutes of viewing. Pay movie channels accounted for 5.8 channel changes and more than an hour of viewing per day. Superstations and independents were each watched four separate times in a day, for about 45 to 50 minutes. General satellite channels accounted for 6.5 changes and about an hour of viewing, spread across about two of the six channels. The remaining channel types each accounted for only 4 to 8 minutes of

Table 2. Household Viewing Averaged Across Days by Channel Type

Channel Type	No. of Channels	No. of Channels Viewed	No of Changes per HH	Viewing Time	% of Time	Expected[a]
Network	4	2.44	10	3:22	44%	12%
Pay Movie	2	1.12	5.8	1:17	17	6
Superstation	3	1.32	4.1	:46	10	9
Independent	4	1.38	4.2	:52	11	12
General Satellite	6	1.76	6.5	:58	12	17
Automated	5	.57	1.3	:08		
PBS	2	.42	.7	:07		
Specialized					6	44
Sattelite	4	.47	1.0	:07		
Access	4	.18	.3	:04		
TOTAL	34	9.66	33.9	7:41	100%	100%

[a]Number of channels of that type as a proportion of total channels available.

viewing per day, and three changes to one to two different channels for all 15 channels.

The proportion of viewing by channel types by day show virtually no variation from day to day. Despite the fact that network stations account for only 12% of available channels, they consistently drew over 40% of the average household's daily viewing time.

Certain other channel types performed substantially better than their proportionate availability. Pay movie channels comprised 6% of the channels but attracted 17% of the viewing time. This is particularly strong in view of the fact that both HBO and TMC were subscribed to by only about three-fourths of the households.

On the other hand, the satellite channels drew 12% of the total viewing time on 17% of the channels. Finally, the remaining 15 channels received only 6% of the viewing time. While this set of channels does include rarely viewed access channels, it also includes the Christian Broadcasting Network and the public television stations, which does not bode well for these stations if their success is to be judged by viewing levels alone.

Stretch viewing by channel type.   Here we move to a special channel-type profile analysis, focusing on stretch viewing. The average stretch lasted 57 minutes (long enough for an hour show and longer than a half-hour show or the shorter segments featured on Cable News Network or MTV).

Three summary statistics which describe the system-wide use of each channel type were calculated: the proportion of channel changes to that type which resulted in stretch viewing, the proportion of time spent viewing that type in stretches, and the length of the average stretch. These data form "portraits" of different "channel usage type" patterns—for instance, a channel with a low proportion of the audience at any one time, but which is sampled relatively often.

Table 3 shows figures for key channel types with medium- to high-viewing rates. The network broadcast stations led in all, which is to be expected as they are also the most watched by about three to one over any one of the other categories. Of interest here, however, is the overall average stretch of 59 minutes. This suggests a tendency to watch in hour blocks.

Pay movie channels' stretch-viewing time proportion and average stretch time are comparable to those obtained by the networks. However, only one in five changes to those channels became a stretch, as opposed to one in three for the

Table 3.   Stretch Viewing by Channel Type

	% Stretch Changes	% Stretch Minutes	Time/Stretch
Networks	32%	88%	:59
Pay Movie Network	20	84	:57
Superstations	16	77	:57
Independents	20	76	:49
Satellite Networks	12	65	:49

networks. This would seem to indicate that there is more sampling of these channels, either to decide if they like the movie or just find out what movie it is. The average stretch of just under 60 minutes for these channels is particularly interesting given their primary content, feature length films which typically last 90 minutes to 2 hours. People do not consistently watch entire movies. A further question would be whether these channels are often used to "fill in" gaps, whether there is a desire for some break or intermission in the film, or a long-term film sampling process over several days.

The six channels clustered under "satellite" show the lowest proportion of changes leading to stretches (one in eight) and the lowest proportion of viewing time spent in stretches (two-thirds).

Because this group of channels displays considerable heterogeneity within itself and across the days studied, some individual channel statistics are of interest. The MTV channel seems to be used in the same manner as a typical radio station: if a preferred song is on, the viewer continues watching (and listening); if not, the viewer tunes out. Only 9% of the changes to MTV led to watching for 15 minutes or more, implying that people do tend to sample the channel and tune out if nothing interesting is on. In contrast, the Cable News Network is more likely turned to for viewing in stretches (one-fifth of the time), and has the largest proportion of its viewing in stretches among the satellite channels. The CNN format of 15- to 30-minute news shows apparently induces that viewing pattern. The sports channels, ESPN and USA, show far more variability from day to day on all these viewing measures, likely due to the inconsistencies of weekday sports programming. Lumping general satellite channels as a single type seems less warranted than the other channel groupings.

A final contrast is that between the superstations and the local independents. Since the superstations are basically independent stations "broadcasting" via satellite, the content across these two categories remains roughly the same. As a result, we might expect similar stretch-viewing patterns, with a slight advantage going to the local independents based on their local orientation and greater past experience with them by the audience. The proportion of changes leading to stretch viewing was larger for the local stations, but the average stretch length was longer for the superstations. This may be because of a greater number of lengthy sports programs on the superstations, specifically (in June) baseball, one baseball game being equivalent to about 3 hours of programming.

Daypart Analysis

The data reported here are exemplary of another approach to cable-viewing analysis, focusing on aggregated viewing behavior by time of day. They also provide an initial examination of *when* most channel changes occur.

Although specific patterns will vary, one generally assumes that most program changes occur on the hour and half-hour. Channel changes that are made at other than those focal points likely reflects changes during a program or changes during commercial time inside a program. The number of channel changes occurring at each minute in a day was calculated, and aggregated into a composite hour of channel changing. All channel changes at exactly 1 minute after the hour, for the 24-hour day, were summed to derive the total number of changes occurring at 1 minute after the hour for a given day. This was done for each minute in a composite hour cycle.

If channel changes were randomly distributed across an hour cycle, the number of changes each minute would be expected to be equivalent to 1/60 of all changes. Tuesday contained 5,651 total changes, Wednesday 5,571, and Thursday 6,121. In order to compare the pattern of changes among the 3 days, the data were normalized to remove differences due to variation in overall number of changes per day. The normalization formula used converted the raw score into

Figure 1.   Daypart Analysis

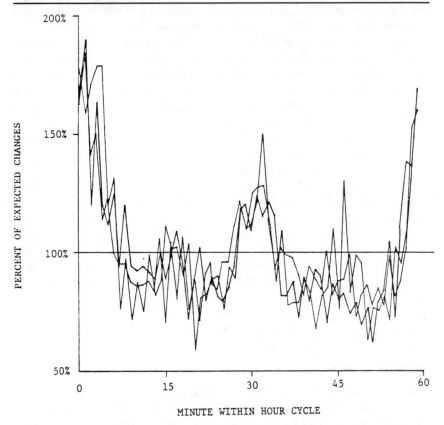

(Percentage of expected channel changes was calculated for each minute in the composite hour cycle by dividing total changes that actually occurred during that minute by 1/60 of the total channel changes for the entire day. Plots of observed channel changes are overlayed for each of the 3 days' aggregated data.)

the percentage of observed changes that would be expected if changes were evenly distributed. Thus, 100% means that exactly 1/60 of the day's changes occurred at that minute in the cycle. Values greater than 100% exceed chance expectations; values less than 100% are less than chance.

Figure 1 contains a graph of the 3 days. Three major statements can be made on the basis of this analysis. First, channel changing does not appear to be random. Definite cycles are observable, with the largest peak between 57 minutes after an hour through 7 minutes after the next hour. The other large peak is at the half-hour. These peaks correspond to time periods between programs. Second, channel changing is by no means confined to time between shows. The number of changes rarely drops below 70% of the changes expected with evenly distributed changing for any minute in the day. There are minor peaks around 15 and 45 minutes (possibly commercial break time between shows), but changing occurs regularly throughout the hour cycle. Finally, the 3 days' patterns are very similar; the cycle repeats itself across days.

DISCUSSION

This exploratory analysis has defined and verified a number of cableviewing behaviors and approaches to interactive audience measurement data. It provides behavioral operationalizations for evaluating self-reported claims of channel changing, "zapping," viewing styles, and viewing choices on cable.

String, stretch, and ministretch viewing behaviors appear regularly, at consistent levels across days. String and ministretch viewing are substantial viewing time components, accounting for nearly one-fifth of total viewing. There is extensive variation among households in the frequency and nature of the different patterns, but little variation in a given household across days, suggesting that households exhibit consistent viewing styles. The viewing-time categories appear to describe meaningful and different behaviors, although their operationalizations of specific time parameters should continue to be evaluated.

The channel profiles show that channels available only on cable successfully drew almost half the viewing time for these sampled households. Viewing of the cable channels is spread across many more channels than off-air viewing, evidencing audience fragmentation both between cable and noncable households and within cable homes. As with all of the other analyses, viewing by channel type was very consistent across days. Identifying the amount of time allocated to a specific channel type, and breaking that time down by proportion of stretch viewing, shows that different channel types are watched in different ways. Without further analysis, it is not known whether these differences are actually based on different groups of viewers, with different viewing patterns, or if the nature of the channel type commands a different approach even among the same set of viewers. Viewership among households that actually select a channel would describe *how* the channel is viewed, rather than how much it is viewed, but viewership overlap among channels also has to be assessed.

The uniform consistency across days suggests the presence of extremely habitual behavior, almost uninfluenced by the day of the week or the program schedule. Duration of the average stretch, per channel, appears to be the major source of variation across days.

Similarly, we have begun to explore the concept of "channel repertoire." The number of different channels viewed, the mix of channels (from the different channel types), and the diversity of viewing patterns allocated to different channels by the same viewer have interesting possibilities for advancing understanding of program selection processes. These findings suggest that a consistent repertoire may exist, both at an aggregate and individual level. The reasons for, and implications of, individual differences bear further examination. The concept of a channel repertoire may lead to a reconceptualization of those types of models which attempt to predict viewership, concentrating on a set of channels to which a viewer is loyal rather than on the current channel to which the set is tuned.

The daypart analysis demonstrates that channel changing on an aggregate level is cyclical. Again, the consistency across days is striking. Although changing peaks occur on the hour and half-hour between programs, advertisers and programmers may be less than delighted to learn that channels are changed at quite high levels throughout the day, during commercial time and during programs. The daypart analysis and the string analysis suggest that zapping (eliminating commercials by quickly changing channels) may not be the dominant motivation for channel changing with cable. Channel changing is distributed across an hour cycle, with the majority of changes occurring other than at half-hour and hour peaks. The average string period is about 5 minutes, and frequently strings last even longer. If zapping were the sole object, presumably strings would last only as long as a commercial break. Why viewers change channels—and why they engage in extended string viewing—remain unanswered but intriguing questions. New dimensions of behavior are now available to characterize the "active viewer."

The data set analyzed here is not without limitations. For instance, data were all collected during a summer week in June, and only weekday data analyzed. What happens to viewing patterns on weekends? How stable is a household's channel sampling repertoire from week to week? There is no guarantee that television viewing in June, while the networks are in reruns and overall viewership is at its lowest level, resembles viewing of first-run network shows. Cable systems are also very diverse, and the behaviors observed may not be generalizable to substantially larger or smaller systems.

From a technical, hardware-oriented standpoint, two questions arise. The headend computer could sample households at 1-minute intervals, but a minute is a relatively long time for analyses concerned with in-depth descriptions of strings, where 5- and 10-second station identifications and 30-second commercials are important. What is lost by this 1-minute limit? Additionally, the cable system studied is equipped with remote control, digital channel changers. Do systems using nonremote devices show the same sort of extensive channel use pattern, or do they show longer stretches and fewer changes? We think there will be a difference, but subsequent research must verify that.

Initial attempts to place such behaviors in a more theoretic framework might begin from different vantage points. A uses and gratification perspective suggest several questions: What motives exist to explain the extensive diversion from network offerings? Are viewers driven away from the networks to fulfill certain gratifications, or are they attracted to alternative offerings? Are these substitute or supplemental means of satisfying information and entertainment needs? Do certain offerings, e.g., MTV, create and gratify needs not previously sought from television?

Decision-making models would have us explain better the process by which choices are made, in an expanding choice system: To what extent and by what means do viewers become aware of the range of choices available? What criteria are used to make program choices? This seems especially to be an iterative decision-making system, in which the process (or processes) employed by the viewer are recycled during a show, between shows, as well as across days. Is the decision process for an individual stable? How much of what the cable system offers is learned and sampled before a particular decision-making pattern becomes embedded in individual viewers? What happens when new entries (channels) are added to the available pool?

Attempts at question-asking and hypothesis-information about cable viewing should address the conceptual elements of a multichannel TV environment. This study examined viewership in a 34 channel cable environment, labeling the behavior "cableviewing." Of course, multiple channels are available to noncable viewers as well, though not nearly as many as with cable. What is the relationship between viewing style and channel availability? It is possible that cable- and broadcast-viewing differences are a matter of degree rather than kind? Parallel research in systems with fewer and more channels could provide a quasi-field experimental test of certain of these notions, combined with equivalent data from broadcast-only households.

The present identification and description of a subset of critical cableviewing behaviors, electronically measured, offer a more precise conception of "television-viewing styles," with rich implications for further study.

SECTION TWO

Viewing Styles

## CHAPTER 5

Profiling the Zappers

Carrie Heeter and Bradley S. Greenberg

Zapping: Is it a myth as some argue, occurring so seldom as to not warrant advertiser attention to a "missing" audience? Or is it a parasite that will grow as more homes acquire cable, VCRs, and remote control? Zapping of commercial content has been a major television industry concern. With the cost of TV advertising time rising, rumors that the viewing audience of commercials is shrinking cause consternation among the stations and networks collecting the revenue and raise objections from the advertisers footing the bills. Zapping commercials (changing channels when commercials come on) was the initial problem. Some have suggested that almost as many viewers "tune-in" as "tune-out" during commercials, so there is no net loss in audience size. "Zipping" has been added to the list of concerns: fast-forwarding through commercials while watching a show recorded on VCR.

There are studies which have yet to be conducted to assess such things as viewer retention of commercial content only partially viewed or viewed in fast forward, what kinds of commercials lead to zapping, whether zappers return to the original channel before commercials are over or miss part of the program content, etc.

This chapter will focus on the issues of *who* zaps commercials and what other viewing behaviors zappers engage in. It will address the question of what segment of the audience tends to be lost to zapping. Data will be drawn from many of the studies reported elsewhere in this book.

Zapping of commercials was most typically assessed in our studies by the question "How often do you change channels when commercials come on during show?" Two related questions which appeared on most surveys were "How often do you change channels during a show, not at commercials?" to tap zapping of program content, and "How often do you watch more then one show at a time by changing channels?" Figure 1 graphs the frequency distributions of responses to the two channel changing questions across two adult studies. One was a door-to-door survey of 230 respondents who subscribe to cable, and one a telephone survey of 339 randomly selected respondents reflecting the 60% cable penetration of that same city. Changing channels during a show, not at commercials, is similarly infrequent in cable and mixed households. (Note, however, that remote channel-selectors were not common in the cable system being studied.) Changing channels at commercials was reported more frequently by the cable sample, with the major difference between 100% cable and 60% cable samples appearing in the ''Almost Always'' category. In other words, the difference greater cable penetration in one sample made was to add 15% of the respondents to the frequent zappers category. It was the extreme zapping behavior that the presence of cable enhanced most. The effect was found for zapping commercials, not zapping program content.

Figure 2 presents the frequency distributions for three zapping questions across 407 fifth graders, 395 tenth graders, 351 fifth grade parents, and 180 tenth grade parents in a 50-channel cable system with 67% penetration. These surveys were conducted 2 years after the adult surveys graphed in Figure 1. The scales are different. While Figure 1 uses a five-point scale ranging from Almost Always to Almost Never, Figure 2 uses a four-point scale from Very Often to Not At All. The distribution shapes are different. In Figure 1, the modal category is Almost

Figure 1.

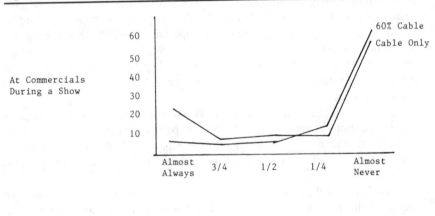

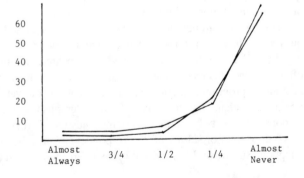

Figure 2.

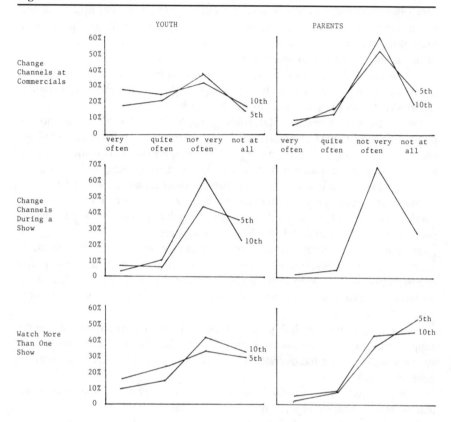

Never, while in Figure 2 the mode is usually Not Very Often. Whether this reflects scaling differences or real differences is uncertain. A further confounding factor is that the responding parents were disproportionately female, and, as we will see later in this chapter, females tend to zap less often than males.

Across all three questions, fifth and tenth grade distributions are similar to each other, distributions for parents of both grades are similar to each other, and parent–child distributions are distinctly different. Changing at commercials during a show was the most commonly engaged in behavior. Fifth and tenth graders reported zapping commercials more frequently than their parents. The distribution shows about one-fourth (19%–29%) of the youth zapping commercials very often, almost another fourth doing so quite often, and the remainder zapping not very often or not at all. Tenth graders were more likely to be extreme zappers than fifth graders by a 10% margin. For parents, less than 10% reported zapping commercials very often, 12%–14% did so quite often, and more than three-fourths (78%) zapped not very often or not at all.

Around one-tenth of the youth say they watch more than one show at a time very often, compared to 5% or less of parents. Sixteen percent to 23% of the youth watch two shows at a time quite often, compared to 8% of the parents. Sixty-three percent to 74% of the youth watch two shows at a time not very often or not at all, compared to more than 85% of parents.

Consistent with the other behaviors, youth change channels during a show more often than parents, although both groups do so infrequently. Tenth graders report slightly more channel changing than fifth graders.

On the basis of the earlier adult studies, where 55%–65% of respondents reported that they zapped commercials "Not At All," viewers were classified dichotomously as zappers and nonzappers for comparison. This distinction was used to examine the question: Who zaps? Across a wide range of demographic variables, two stand out as fairly consistently discriminators: sex and age.

Males are more likely than females to report zapping commercials. In some studies, they do so by nearly a 2–1 margin. Sixty-five percent of zappers in one study were males, compared to 37% of nonzappers.

In terms of age, young adults zap more than older adults. Average adult zappers are in their 30s, and nonzappers are in their mid-40s. One could speculate that those who grew up with television do more channel changing than those for whom television was a later-life phenomenon. And we have evidence to support that speculation.

Our study of fifth and sixth graders shows that they report significantly more zapping than do their parents (who are in their mid-30s). Projecting these data strongly suggests that the overall level of zapping will increase as the younger generation matures.

Zappers have not been found in these studies to differ by income, education, marital status, household size, or number of children.

Furthermore, zappers do not own more technology toys than nonzappers. Zappers and nonzappers equivalently purchase VCRs, videodiscs, video games, and home computers.

However, zappers do differ in the extent to which they have remote-control channel selectors. Thirty-four percent of zappers, compared to 27% of nonzappers, had remote control. Although remote devices may be more plentiful in cable settings, it is the device, more than having cable, which is related to zapping commercials. Getting up to make changes at the TV set remains a prime inhibitor of zapping.

How do zappers watch TV? First, let us look at how much prior planning goes into television viewing before approaching the set. Zappers do less planning. In every study we have done, zappers are less likely to know what shows they are going to watch before they sit down to watch. They are less likely to check a television guide, and they are less likely to look at the week's offerings to make advance viewing decisions. They are more likely creatures of chance than of habit. Some studies show that zappers are less likely to watch the same shows from week to week, less likely to watch the same shows that are stripped daily, and more likely to watch shows they do not watch regularly. This applies to child zappers as well.

Given the lack of advance planning, it is not surprising that zappers also behave quite differently when they do turn on the television set. Their orienting searches are characterized by a good deal more searching. Among these behaviors, the following are worth noting. Zappers are more likely to change channels before deciding what to watch. They check more channels when deciding what to watch. They are more likely to check channels in numerical order (2,3,4,5,6, . . .36). They are more likely to stop at the first show that looks good to them. Zappers, then, are channel-checkers when they turn the set on.

Having settled into a television show, viewers have the continuous option of reevaluating their choice. Zapping commercials is one type of reevaluation. What reevaluation behaviors co-occur with the zapping of commercials? They are numerous. Zappers exhibit more reevaluation behaviors. Zappers are less likely to watch a show from start to finish. They are more likely to change channels between shows, during shows, and at commercials. They change substantially more often during shows they regularly watch and during other shows. During regular shows, zappers change channels an estimated 6.8 times per hour, compared to 0.8 time for nonzappers. During nonregular shows, zappers change an average of 9.2 times per hour, compared to 2.0 times for nonzappers.

From this set of consistent behaviors, the generalization is evident: Zappers zap more of everything. It should be noted that zappers check guides more often while viewing than nonzappers. What they do not do is use guides to plan ahead.

Across all of the studies in which we asked the question, zappers were more likely to watch and follow more than one show at a time. This is a new and interesting phenomenon for communication researchers. An efficient channel changer may easily follow two storylines concurrently and is even more prone to use commercial time to do just that if the commercial segments on two or more shows are offset in time. We plan to do experimental work in the future involving the watching of two, three, or more shows concurrently to assess learning, satisfaction, attention, commercial exposure, and so on. New definitions of what it means to be a program viewer may need to be developed. There is ample evidence that viewers do not watch and do not need to watch an entire show to claim to have seen the show, perhaps, independently of their allocation of real time to the show.

Furthermore, this set of reevaluation behaviors passes from parent to child (or perhaps child to parent). In our study of fifth and sixth graders and their parents, children of zappers zapped more and reevaluated more than children of nonzappers. For example, checking to see what else is on during shows, both within and outside of commercial time, and checking elsewhere at the end of show were all significantly correlated between parent and child.

Why do zappers zap? In one study, we sought the motives for frequent channel changing. The primary reasons for channel changing were to see what else was on and to avoid commercials (see Figure 3). The largest difference in motives between zappers and nonzappers was to avoid commercials. Boredom, variety seeking, and multiple-show viewing were also more often cited by zappers. Many of these reasons may be as much program oriented as commercial bound.

Together with this channel-changing syndrome, zappers also consistently tell us that they pay less attention to television in general. Furthermore, zappers use television more for background sound than nonzappers, and they more often have the television on without audio. These suggest that strong zappers also use television more as an accompaniment to other activities in the household. However, there was no significant difference in total television viewing time between zappers and nonzappers.

Zappers use more channels. They are familiar with more different channels— four or five more, on the average, in a 36-channel cable system. Zappers also reported larger channel repertoires of regularly viewed channels, three additional stations above and beyond those watched regularly by nonzappers. Given no difference in total viewing time, then, zappers spread themselves somewhat more thinly across the options available to them.

These partial findings from our studies feed the next generation of research questions: Can zapping be inhibited, or channeled? Perhaps one reason for the amount of searching that occurs is inadequate packaging of program information in systems with dozens of channels. Packaging of commercials may also need to be addressed. In our baseline survey for the system-specific guide study, we have found that zappers and nonzappers look for the same kinds of information from a guide, but we have yet to determine whether they are satisfied with what they find.

Is zapping habitual or responsive to content? We do not know the extent to which commercial zapping is sheer avoidance of commercial messages or a response to the first in a string of commercials that is aversive to the viewer. We do not know whether reevaluation of a chosen program (checking other options in

Figure 3. Why Zap?

2.4	To see what else is on
2.1	To avoid commercials
1.8	Because they're bored
1.5	For variety
0.9	To watch multiple shows

4 = almost always, 0 = almost never

midstream) is just a way of watching television or is a response to aversive program content or program types. No study has yet determined, for example, the extent to which the zapping of one commercial leads to viewing of an alternative commercial. Examination of program content when zapping occurs and experimental manipulation of commercial stimuli would be useful in distinguishing between viewing habits and selective avoidance.

Although content factors likely affect channel changing, zapping appears to make sense when considered in the context of the viewing style we have outlined. It is not an idiosyncratic behavior aimed only at television commercials. It is an element of a systematic set of behaviors that fit together as an approach to watching television. This argues for studying commercial zapping as part of a broader orientation to television. It is akin to finding a symptom and wanting to understand exactly what the problem is or if there is a problem. We suggest that there may be several typologies of television-viewing styles—ranging perhaps from the channel-loyal, total zap abstainer to the superzapper for whom television viewing may be a sporting event (i.e., how little time can I spend watching and still get as much as I want from all the choices available?).

Is zapping growing? We expect that to be the case for both technological and developmental reasons. More viewers will have some form of remote-control capability; more viewers will have VCRs and use them as a personal editor; child viewers will have matured with multichannel systems, and changing channels will be as natural to them as nonchanging is to their grandparents.

## CHAPTER 6

New Fall Season Viewing

Carrie Heeter

INTRODUCTION

This fall, like every fall, a new season of shows was introduced by the networks: the latest offerings in an unending competition to attract viewers. The fall season network lineup presents viewers with an accentuated program choice situation. Summer reruns end, and old and new series vie for viewer attention. This period is crucial to the viability of new shows, and the continuance of returning shows. During the first weeks of a new season, the viewer might be expected to be most acutely aware of alternatives to watch.

National ratings services report audience viewing statistics (the *outcomes* of the choice process). But the actual behavior of individual viewers as they approach a new season has yet to be examined. How do viewers respond to these new alternatives? Do they sample program options and select favorites for future regular viewing? How quickly does the sampling cease, and viewing patterns stabilize? Is it a period of major change in viewing habits and channels watched? Is the impact limited to network channels with competing fall lineups, or does it become a period of general channel evaluation?

Viewership data collected via two-way cable provide an opportunity for in-depth examination of viewing style and channel usage among cable households. This book presents two studies of the 1982 fall season. Chapter 6 examines viewing style and audience reactions to new season prime-time offerings. Chapter 7 is a subanalysis of what channels cable households with children watch on Saturday mornings, and how they watch those channels. The studies share a common methodology for data collection. They are based on the same or overlapping sets of households, at approximately the same time period.

The overarching issue in this chapter is: What is the impact of a new season of network shows on cable subscribers' viewing styles? The network broadcast environment for summer and fall are quite different. (See Figure 1.) In the summer, ABC, CBS, and NBC carry reruns and invest in less program promotion than the rest of the year. In early fall, intense promotion and special pilot programs herald

Figure 1.   Hypotheses

BROADCAST ENVIRONMENT

	NETWORK PROGRAMMING	NETWORK HYPE
SUMMER	reruns	low
NEW FALL	top quality	peak
FALL	high quality	high

VIEWING TIME

	(H_1) VIEWING TIME	(H_2) NETWORK SHARE
SUMMER	(low)	(less to nets)
NEW FALL	high	more to nets
FALL	higher	most to nets

VIEWING STYLE

	(H_3) CHANNEL SAMPLING	(H_3) CHANGES/ HOUR	(H_4) NETWORK SHARE	(H_5) CHANNEL REPERTOIRE
SUMMER	(high)	(high)	(lower)	(highest)
NEW FALL	higher	highest	higher	lower
. ALL	lower	lower	high	higher

the new season, yielding next to normal in-season-quality programming and hype.

Audience viewing time and network share of viewing time have been examined over this change in broadcast environment by other research. The average time viewers spend each day watching TV is highest in mid-winter and lowest in mid-summer. Gensch and Shaman (1980) suggest that viewership is as much a function of amount of available daylight as it is relating to program preferences. Their trigonomic regression model using only time of the year as a predictor was able to account for 85% of the variance in viewing time at 7:30 p.m., over 1 year. The introduction of new season programs in fall may increase viewing time slightly, but it does not alter that basic patterns. This study will examine changes in overall viewing time among the sample cable subscribers.

> H1: Viewing time will increase as the fall season continues, rather than reach a peak during the first weeks of the new season.

Even if viewing time is not particularly responsive to program content or promotion, the network share of viewing time should increase in early fall. Studies have detailed how viewing is allocated in broadcast, basic cable, and pay cable households. Weekly shares for local network affiliates and public television stations are lower in households with cable (Webster, 1983; LeRoy & LeRoy, 1983). Local stations maintain 40%–80% shares, depending on market size. Pay cable channels attract a 14%–20% share in pay cable households across different market types. Distant stations draw a 10% share in the largest markets, and as much as 46% in small markets. The other cable-only channels reportedly account for very little viewing time—1%–7%, with an average across eight market types, including pay and basic households, of only 3.2% (Webster, 1983). These averages derive from secondary analysis of Arbitron diary data, which are known to underreport local independent, imported broadcast, basic cable, and pay cable channels, and to slightly overreport local network affiliate and public television viewership ("Arbitron Ratings," 1983).

A sudden but gradual increase is the pattern for network share of viewing time in early fall. Weekly Nielsen network share ratings for fall of 1984 were as follow

August		September		October	
Week 1	75.7	Week 1	74.8	Week 1	79.2
Week 2	75.7	Week 2	74.8	Week 2	80.1*
Week 3	75.7	Week 3	76.4	Week 3	79.6
Week 4	75.6	Week 4	77.8	Week 4	79.2

*Indicates peak network viewing share for the year

Beginning the third week of September, network share increased gradually to the season high of 80.1 in the second week of October. Then, network share dropped back. Thus, the new season introduction appears to have affected network share. The two-way data will permit detailed examination of observed rather than self-reported cable household use of the various channels. In addition, changes across the new season can be measured.

> H2: Network share will be lowest in summer and highest in early fall as the new season begins.

Research attention has also turned to differences in *how* people watch television. A well documented difference in cable household viewing style is increased channel changing and channel sampling, compared to broadcast households.

Chapter 4 identified behavioral components intended to reflect different viewing styles, using interactive viewing data. Three modes of viewing were identified: *channel sampling*, where a short period of time (0–4 minutes) is spent with a single channel; *program viewing*, where a channel is watched for an entire program (approximated at 15 minutes or more); and *extended sampling*, which falls between short sampling and program viewing, where a program is watched long enough to be exposed to some program, but is watched for less than the entire show (5–14 minutes). Channel sampling accounted for an average of 8% of viewing time, extended sampling 10%, and program viewing 82%. When viewership to a given channel or channel type is analyzed using these measures, a profile emerges of how channels are used, in addition to how much.

As a fall season begins, viewers are faced with a new menu of shows (some returning, some being introduced for the first time), and promotional efforts accentuate awareness of the choices. During this period, sampling different new programs might be expected to increase, peaking during the early weeks and leveling off as the season continues and viewing choices are made.

Similarly, if viewers are aware of several potentially appealing program options, they may be more likely to turn away from a program in midstream, either out of curiosity about what other shows are like, or because the quality of the show they initially selected did not meet their expectations. Thus, just not channel sampling but channel changing at all points in a show may peak at the beginning of the season.

H3:   Channel changing will be most extreme in early fall.

In summer, 30% of all channel changes and 44% of viewing time were to network stations (see Chapter 4). The network share of viewing time increases in fall. What happens to the network share of channel changes? If there is increased channel changing for the reasons specified thus far, then it should be increased channel changing to network channels, in response to programming and promotional changes.

H4:   The network share of channel changes will be highest in early fall and lowest in summer.

If the hypothesized increase in channel sampling is found, but no increase in network share of channel changes, then perhaps the new season produces an atmosphere of increased sampling and selectivity which generalizes across all channels.

Various studies reported in this book and elsewhere have measured channel repertoire (the set of channels viewed regularly by an individual or household). In summer, the household channel repertoires in this cable system consisted of an average of 2.4 network stations and 7.2 other stations, with a network share of 44% of viewing time. Presumably, in fall, the networks draw a larger share of viewing time, at the expense of cable-only and other channels. Channel repertoires may also become more focused on the networks, such that some nonnetwork channels watched in the summer are not merely watched for less time in fall, but are not watched at all in some households. The early season may be a transitional period when channel repertoire is changing, getting smaller and more exclusively including network channels as fall continues.

> H5:   Channel repertoire will become smaller as the fall season progresses. If the
> channels checked do not change amidst changes in viewing time and
> viewing style, then the new season may primarily provide for reallocation
> of viewing time across a fixed set of channels regularly viewed. (See Figure
> 1 for a summary of hypotheses.)

Methodology common to the two fall season studies is described next. Chapter 7 begins with its own issue-specific introduction and brief methodology section describing how child households were identified.

METHODS

Data Collection

Data were collected in a 35-channel interactive cable system in Florida the first, second, fourth, and fifth weeks of the 1982 fall season, during prime time and on Saturday mornings. The initial selection of the sample 184 households was random, but substitution of adjacent locations was permitted where necessary. The headend computer in the cable system scanned viewing behavior in the sample households every 20 seconds, storing *time* of each change status (either on/off or changes to a different channel), the *channel changed to*, and *the household identification number* whenever a change in status occurred.

Three types of errors were possible: *system-wide error*, affecting all data collected; *area-wide error*, affecting all households served by a given "code operated switch" which controls the upstream data flow for that geographic sector; and *household error*, reflecting technical problems with a particular home terminal. The monitoring system was programmed to recognize and monitor these errors, recording error condition and time of day when they occurred, as well as noting when they were corrected.

Ten sample days were discarded due to high area or system wide errors. (The system was in an experimental stage. Fifteen household terminals consistently registered malfunctions, and another four to 29 terminals registered substantial errors—more than half of the viewing time monitored on some days.) On a day-by-day basis, households with more than 50% error time were treated as missing data. Eleven households were not activated until partway into the first week of data collection. The resulting usable household sample sizes ranged from 139 to 165 households, monitored on a given day, with an average error rate within

Table 1. Error Rates

Date	# Households Surveyed[1]	# Viewing Households[2]	Percentage of Error Time[3]
WEEKDAYS			
Week 1			
9/27	161	120	1.5%
9/29	155	123	0.8%
9/30	165	120	1.1%
Week 2			
10/6	139	113	0.6%
10/8	151	126	0.9%
Week 4–5			
10/21	165	129	0.6%
10/22	164	117	1.0%
10/25	164	140	1.0%
SATURDAY MORNINGS			
10/2	150	80	2.0%
10/9	165	85	1.3%
10/23	164	96	1.0%

[1]—number of wired terminals with acceptable error rates (less than 100 minutes out of 180 for prime time or 240 for Saturday morning)
[2]—number of viewing households with acceptable error rates
[3]—percentage of time measured in viewing households which registered as errors

those households of 2% or less of total data collection time. Table 1 reports specific sample sizes and error rates, by day.

Predominantly error-free data were thus obtained for the following days:

	M	T	W	TH	F	SA
September	(27)	28	(29)	(30)	1	(2)
October	4	5	(6)	7	(8)	(9)
	11	12	13	14	15	16
	18	19	20	(21)	(22)	(23)
	(25)	26	27			

◯—indicates sample day

Sample Description

In addition to the interactive viewing data, the households were surveyed. Mail questionnaire was used, with overt household identification (see Chapter 10). Those households which did not return a questionnaire within 10 days were interviewed by phone. Surveying occurred approximately 6 months after collection of viewing data, to avoid sensitizing respondents when the system was being used for data collection. One hundred and fifty of the 200 sample homes completed surveys: 39% were by telephone, 61% by mail. The average age of the adult respondent was 47. Mean education was 2 years of college. Average household income fell between $30,000 and $40,000. These demographics suggest caution in generalizing the data to the entire population of cable subscribers due to upscale income and education. Nevertheless, the data do provide an opportunity for an initial detailed examination of how households use cable.

Description of Cable System

The Florida cable system carried 35 channels, including two pay services (HBO and The Movie Channel, with 150% pay penetration), four network stations (one a distant ABC affiliate), four independents, three superstations, two PBS stations, four local access channels, and 16 other cable-only channels. Table 6 contains a complete listing of available channels.

Variable Construction

Fortran programs were written to create specific viewership and viewing style variables from the raw viewership data. Each variable created described a single day's household viewing pattern during the specified measurement period. For weekdays, the period was prime time (7:00 p.m. until 10:00 p.m.); for Saturdays, it was mornings from 8:00 a.m. until noon.

Viewership.   Across households, the amount of viewing time and share (proportion) of viewing time allocated to each individual channel and type of channel were calculated. Throughout this report, *network* viewing includes the three local network stations plus the distant ABC affiliate. *Other off-air* viewing combines two local PBS and four independent stations. *Pay cable* groups TMC and HBO, and *other cable*, includes the remaining 23 cable-only channels.

The number of channel changes, and proportion of channel changes to individual channels and to types of channels, were also determined. For example, the average number of times a household changed channels to CBS, and the proportion of all channel changes CBS accounted for, were calculated.

Viewing style.   For individual channels, two measures of circulation (cume) were created, using these criteria: (a) the percentage of households that watched or checked the channel *at all*, and (b) the percentage of households that watched the channel *for 5 composite minutes or more*. Two parallel measures of channel repertoire were created: (a) total channels watched or checked by a household, and (b) total channels watched for 5 composite minutes or more.

Other measures of channel switching follow those developed in Chapter 4, but program viewing was further differentiated into "stretches" of 14 to 24 minutes watching a single channel, and "maxi-stretches" of 25+ minutes. This precision was added to enhance the ability to identify changes in viewing style over time. Channel sampling was viewing any single channel for 4 minutes or less, and extended sampling referred to 5 to 14 minutes on a single channel. Thus there four ways to watch a channel: sampling, extended sampling, stretch, and maxi-stretch. The average number of channel changes per hour overall, and to different types of channels, were also computed.

Comparison Data

Summer data used for comparison are drawn from the earlier study of this same cable system, including approximately 75% of the same households (Chapter 4). The summer data did not separate prime-time viewing, and instead measured the entire day's viewing across 3 days. Some measures were sufficiently incomparable that they were omitted. Week 1 results report average findings for three weekdays; Week 2 for two weekdays; and Week 4–5 for three weekdays (Thursday and Friday of Week 4, and Monday of Week 5).

Results

Hypothesized results (see Figure 2) are presented first, followed by general observations.

H1:   Viewing time will increase progressively.

Figure 2. Results

VIEWING TIME

	VIEWING TIME	NETWORK SHARE
SUMMER	--	44.0%
WEEK 1	1:44	52.6%
WEEK 2	2:04	52.5%
WEEK 4–5	2:06	59.6%

VIEWING STYLE

	CHANNEL SAMPLING	CHANGES/ HOUR	NETWORK SHARE	CHANNEL REPERTOIRE
SUMMER	--	4.3	--	--
WEEK 1	32.4%	6.4	35.8%	5.2
WEEK 2	27.5%	5.9	35.2%	5.8
WEEK 4–5	24.0%	4.9	36.6%	5.2

Prime time viewing levels were 20 minutes lower in the first week of the new season (1:44) than they were in Week 2 (2:04) or Week 4–5 (2:06). A comparable summer measure from this cable system was not available. While these and all data reported here are more suggestive than definitive, support is provided for Hypothesis 1. No peak is evident, in either Week 1 or Week 2. Viewing time was at the same level in Week 4–5 as it was in Week 2, and Week 1 viewing time was lower than the weeks which followed.

H2:   Network share will be lowest in summer and highest in early fall.

There appears to be a gradual increase in network viewing share, from 44% in summer (overall, not just prime time) to 53% in Weeks 1 and 2, to 60% in week 4–5. Hypothesis 2 is not supported. It appears that network share of viewing does not immediately peak in response to new programming and promotion, but instead that previous patterns change gradually.

H3:   Channel changing will be most extreme in early fall.

In the first week of the new season, during prime time, about one-third (32%) of viewing time was devoted to watching channels for less than 25 minutes in a stretch. This level decreased gradually, to 28% in Week 2 and 24% in Week 4–5. The average number of channel changes per viewing hour can be compared with summer data. In the summer there were an average of 4.3 changes per hour. In Week 1 of the new season, channel changing averaged 6.4 per hour, dropping to 5.9 in Week 2 and 4.9 in Week 4–5. Hypothesis 3 is supported.

H4:   Network share of channel changes will be highest in early fall.

The network share of channel changes for prime time summer viewing is not available. Little variation is evident among Weeks 1, 2, and 4–5. Network share of channel changes remains at 35%–37% during that period. Hypothesis 4 is not supported.

H5:   Channel repertoire will decrease as the season progresses.

In Week 1, households watched or checked an average of 5.2 different channels during prime time. In Week 2, the average was 5.8, and in Week 4–5 the average returned to 5.2. There is insufficient evidence to even suggest a systematic trend. Hypothesis 5 is not supported.

Additional Findings

Viewing share.   Table 2 presents the share of viewing time for broadcast networks, other off-air channels, pay cable channels, and other cable channels. Table 3 identifies changes in individual channel viewing shares across the sample weeks, grouped by channel type.

Table 2. Prime Time Trend Analysis: Allocation of Viewing by Channel Type

	Summer	Week 1	Week 2	Week 4–5
(4) Network				
% time	44%	52.6%	52.5%	59.6%
% changes		35.8	35.2	36.6
(6) Other Off-Air				
% time	13%	13.4%	10.9%	7.6%
% changes		12.6	12.8	10.9
(2) Pay Cable				
% time	17%	10.8%	18.3%	12.6%
% changes		14.2	19.0	16.0
(23) Other Cable				
% time	23%	23.2%	18.3%	20.2%
% changes		37.4	33.0	36.5
# Week Days				
Averaged	3	3	2	3

NOTE: Summer calculations are based on 3 full June weekdays of viewing, not just prime time.

The network's gain in viewing share appears to be primarily at the expense of other off-air viewership, which decreased as the season progressed. Viewing of cable-only channels (including pay cable) changed from 34% in Week 1 to 36.6% in Week 2, to 32.8% in Week 4–5. Pay-cable viewing was highly variable, ranging from 10.8% in Week 1 to 18% to 23% of all viewing, with the level inversely related to the proportion of pay-cable viewership.

Cable networks which consistently received more than 1% of viewing share were: USA Network, MTV, ESPN, and, on occasion, CNN, in addition to the three superstations.

Channel sampling.   Table 4 presents viewing-style trends. Sampling (0–4 minutes), extended sampling (5–14 minutes) and stretches (15–24 minutes), the three indicators of short periods of time spent on a channel, all showed decreased viewing proportions as the fall season continued. Channel sampling dropped from 9.3% of viewing time in Week 1 to 7.1% in Week 4–5; extended sampling dropped from 14.7 to 10.1%; stretches from 8.2% to 6.8%; and maxi-stretches increased from 67.7% to 76%.

For the summer levels reported, maxi-stretch and stretch viewing are combined. Comparing Week 4–5 with the summer period, the late fall percentage was within .8% of summer percentages, suggesting that the heightened channel changing at the onset of the season may have narrowed off to base levels by Week 4–5. Similarly, overall channel changes per hour were lowest in summer (4.3), highest the first week of the new season (6.4), and dropped steadily to 4.9 by Week 4–5.

Table 3.   Individual Channel Viewership

Channel	Week 1 Share	Week 2 Share	Week 4–5 Share
NETWORKS			
9 ABC	15.0	15.9	12.6
12 CBS	23.2	20.6	27.7
7 NBC	13.4	15.1	18.6
18 ABC	01.4	00.9	00.7
OTHER OFF-AIR			
16 PBS	00.7	01.0	00.7
21 PBS	01.5	02.1	01.8
19 Independent	05.5	03.8	01.6
22 Independent	00.3	00.3	00.8
28 Independent	04.0	02.8	02.3
34 Independent	01.4	01.0	00.3
PAY CABLE			
2 HBO	06.2	10.5	06.6
14 TMC	04.7	07.9	06.0
SUPERSTATIONS			
11 WOR	03.3	02.9	02.8
20 WGN	03.9	02.7	01.8
17 WTBS	02.0	01.7	02.2
TEXT			
10 Program Guide	00.0	00.7	00.2
13 Weather	00.0	00.2	00.2
24 Cable Shopper	00.0	00.0	00.1
25 Newspaper	00.0	00.0	00.0
8 Stock Channel	00.4	00.1	00.3
OTHER CABLE			
15 USA Network	02.3	01.9	02.3
23 CBN	00.8	00.1	00.2
26 MTV	02.1	02.0	01.9
27 ESPN	03.4	01.8	04.3
29 CNN	01.3	00.6	01.0
30 SNC	00.1	00.4	00.0
31 SPN	00.3	00.1	00.1
32 Nickelodeon	00.3	00.4	00.1
33 PTL	00.3	00.4	00.4
35 C-Span	00.4	00.0	00.0
36 SIN	00.0	00.5	00.0
ACCESS			
3 government	00.0	00.1	00.0
4 local origination	00.0	00.0	00.1
5 educational	00.2	00.4	00.2
6 public access	00.0	00.0	00.0

Figures indicate share of viewing time to each individual channel, and may not add to 100% due to rounding. For more complete descriptions of the channels available, see Appendix A.

Channel changes per hour.   The number of channel changes per hour was also analyzed by channel type. In terms of overall viewing style of the different types of channels, Table 5 shows that network stations were watched the longest

Table 4.   Prime Time Trend Analysis: Viewing Style

	Summer	Week 1	Week 2	Week 4–5
Channel Sampling (0–4 minutes)	7.9%	9.3%	8.6%	7.1%
Extended Sampling (5–14 minutes)	9.9	14.7	11.6	10.1
Stretch Viewing (15–24 minutes)	82.0%	8.2	7.2	6.8
Maxi-Stretches (25+ minutes)		67.6	72.5	76.0
# Different Channels Checked	—	5.2	5.8	5.2
# Different Channels Watched 5+ Minutes	—	2.4	2.6	2.6
Viewing Time	—	1:44	2:04	2:06
Changes/Hour	4.3	6.4	5.9	4.9
N	3 days	3 days	2 days	3 days

without changing channels (three to four changes per viewing hour). Other cable channels were changed to or from with greatest frequency (an average of nine or ten changes an hour, when watching other cable channels). Weekly variation in changes per hour was different for different types of channels. The average of 4.2 channel changes per network viewing hour in Week 1 dropped to 3.7 in Week 2 and to 2.9 in Weeks 4 and 5. Channel changes to pay-cable channels decreased, from 8.4 to 6.2 to 5.9. For other off-air and other cable channels, the channel changing shares did not demonstrate a consistent trend. As reported earlier, there was an overall decrease in the number of channel changes per hour, from 6.4 in Week 1 to 4.9 in Weeks 4 and 5.

Share of channel changes.   Table 2 contained the share of channel changes for broadcast networks, other off-air channels, pay cable channels, and other cable channels. Table 6 identifies changes in individual channel-changing shares across the sample weeks, grouped by channel type.

The share of channel changes to different channel types was less variable across weeks than share of viewing time. Network channels were watched for slightly more than half of all viewing time, but accounted for about one-third of

Table 5.   Prime Time Trend Analysis: Channel Changes per Viewing Hour by Channel Type

	Week 1	Week 2	Week 4–5
Network	4.2	3.7	2.9
Other Off-Air	5.6	7.1	6.5
Pay Cable	8.4	6.2	5.9
Other Cable	10.3	10.9	9.3
OVERALL	6.4	5.9	4.9

Table 6. Individual Channel Changing

Channel	Week 1 Share	Week 2 Share	Week 4–5 Share
NETWORKS			
9 ABC	09.8	12.7	10.3
12 CBS	14.1	11.4	13.9
7 NBC	09.9	09.3	10.6
18 ABC	02.0	01.9	01.9
OTHER OFF-AIR			
16 PBS	01.8	00.8	01.2
21 PBS	01.8	02.0	01.4
19 Independent	03.6	04.1	03.2
22 Independent	00.8	00.8	00.8
28 Independent	03.7	03.6	03.3
34 Independent	01.1	01.6	01.1
PAY CABLE			
2 HBO	06.7	09.3	07.8
14 TMC	07.5	09.8	08.2
SUPERSTATIONS			
11 WOR	03.9	03.9	03.9
20 WGN	04.0	03.7	03.4
17 WTBS	05.6	04.5	03.3
TEXT			
10 Program Guide	00.7	01.1	01.0
13 Weather	00.1	00.3	00.4
24 Cable Shopper	00.3	00.5	00.3
25 Newspaper	00.2	00.2	00.3
8 Stock Channel	00.4	00.3	00.4
OTHER CABLE			
15 USA Network	03.9	01.6	03.6
23 CBN	00.9	00.8	00.7
26 MTV	05.3	04.5	05.2
27 ESPN	04.9	03.4	05.6
29 CNN	01.2	01.6	01.9
30 SNC	00.6	00.9	00.8
31 SPN	00.9	01.0	00.8
32 Nickelodeon	00.6	00.3	00.6
33 PTL	00.5	00.3	00.7
35 C-Span	00.3	00.2	00.2
36 SIN	00.4	00.4	00.3
ACCESS			
3 government	00.0	00.1	00.1
4 local origination	00.1	00.2	00.1
5 educational	00.5	00.2	00.9
6 public access	00.3	00.1	00.3

Figures indicate share of channel changes to each individual channel, and may not add to 100% due to rounding. For more complete descriptions of the channels available, see Appendix A.

all channel changes, while other cable channels were watched for about one-fifth of all viewing time, but accounted for about one-third of channel changes. Thus, cable channels may be checked more than they are watched, while network channels are watched more than checked. Other off-air and pay-cable channels received viewing and changing shares which were more similar.

As far as individual channels, MTV and ESPN draw 3%–5% of channel changes, as do the superstations and the two most popular independent stations. Pay cable channels range from 7%–10% of channel changes, and networks from 9%–14%.

Channel repertoire.   (See Table 4.) On the average, households checked the approximately same number of different channels at the beginning of the season (5.2) as later in the season (5.8), and watched about the same number of different channels for 5 minutes or more (2.4–2.6) during prime time each fall week, despite differences in the frequency with which they changed channels. The actual channels watched may not have changed, so much as the manner in which, and the amount, they were watched.

CONCLUSIONS

This study has numerous limitations, centering around the relatively small sample size and limited number of days for which data were available, from a single cable system. However, the data offer a unique opportunity to examine actual household viewing behavior across a crucial 5-week period. Here is the tentative picture of new season viewing behavior that emerges from this analysis of two-way cable data.

1) Viewing time increased gradually in fall, and was thus not particularly responsive to promotion or program quality.
2) Network share of viewing time increased sharply and steadily as the season begins.
3) In addition to the three networks and two pay-cable channels, two local independents, the three superstations, and four cable networks (USA Network, MTV, ESPN, and, to a lesser extent, CNN) consistently received 1% or more viewing shares in fall.
4) There was more channel changing early in the season than later. This is demonstrated by the drop in average number of channel changes per hour, from 6.4 to 4.9.
5) Similarly, viewing style changed as the season progressed. At the beginning of the season, viewers engaged in more frequent periods of channel sampling, and spent an average of 68% of their time watching in blocks of 25 minutes or more. By the fourth and fifth week of the season, lengthy viewing blocks accounted for about three-fourths of all viewing.
6) The share of channel changes each type of channel received remained consistent across the new season. Networks received a much greater share of viewing than of channel changes, and nonpay cable channels had the opposite pattern.

7) There were more channel changes per hour to and away from network channels early in the season than later.
8) Despite the increase in channel sampling, households seemed to sample essentially the same number of different channels throughout the new season. What differed was the frequency and duration of the sampling.

The fall season is a period of increased channel sampling, not just to networks but across all channels. Individuals may or may not step beyond the set of channels they usually watch, but the way in which they allocate their viewing time among those channels changes. Network share increases as channel changing decreases.

## CHAPTER 7

Watching Saturday Morning Television

Carrie Heeter

INTRODUCTION

Little research attention has been directed at what children do with cable television, despite vast numbers of studies on children and broadcast television. Only one previous study (Kerkman, Wright, Huston, Rice, & Bremer, 1983) was identified which addressed the topic at all. The authors compared preschoolers' general television environments (TV rules, media orientation, TV availability, etc.) across pay cable, basic cable, and noncable households. The study had two major flaws: sample size was very small, and the cable system in the community studied was only a 12-channel system, no longer representative of the majority of cable systems.

In this chapter, two-way cable viewing data will be used to address two major issues related to children and cable. First, *what* do children watch, given the expanded channel offerings available with cable? A few cable satellite networks are targeted for children (for example, Nickelodeon and the Disney Channel), and several others include programs for children on a regular basis (for example, HBO or USA network) (Siemicki, Atkin, Greenberg & Baldwin, 1987). Do children watch the programming targeted for them? Chapter 6 reported that network and off-air channels comprise approximately two-thirds of all viewing during prime time. How does that level of general prime time network viewing compare to children's Saturday-morning behavior? Do child viewing patterns change over time as the season progresses?

Secondly, *how* do children watch cable TV—what viewing style do they develop to accommodate cable's multiple channels? Will they, like households in general, sample different channels on a regular basis? Will they do so more than, less than, or with the same frequency as, households in general? Do children's viewing styles change over time across the new season, like households in general?

There is some research evidence from noncable studies to suggest that children's viewing patterns include extensive sampling. Lyle and Hoffman (1972) describe viewing behavior of first, sixth, and tenth graders, reporting a strong tendency not to preplan viewing. At all age levels, they found that "students were most likely to turn the set on and flip channels to see if there was a program that interested them" (p. 177). Only about one-third checked a TV guide.

Banks (1980) gathered publicly released Nielsen data on children's viewership, using it to argue that children are no less loyal viewers than adults, and are more likely to "rotate viewing among different programs within a given time period" (p. 52). He cites seasonal and week-to-week variations in program ratings as evidence, noting that ratings among younger children (2–5 years old) showed even less stability than 6- to 11-year-olds. Banks considers the patterns of varied viewing by children to be "an expression of deliberate search on their parts even at quite early an age" (p. 52).

Wakshlag and Greenberg (1979) reach the opposite conclusion in their study of the stability of youth program preferences across a season. Sixth and eighth graders' program viewership was assessed by survey in the fall, winter, and spring. Examining the impacts of various programming strategies on the child ratings, only program familiarity and program start time emerged as significant predictors of regular viewing. None of the other programming strategies had observable systematic effects. The authors conclude that "children are not adventurous viewers, but prefer programs which they know to be entertaining from past experience" (p. 68). Their conclusion may or may not be warranted. An alternative explanation could be that the programming strategies are not effective with children.

Data from a study of parent and child commercial zapping will be reported in Chapter 11. Preliminary analyses found that child viewers report changing channels more often than their parents report changing channels. Whether this difference is an artifact of self-report measures or reflects an actual behavioral difference can begin to be assessed using two-way cable data. This study will attempt to provide initial evidence about viewing style and channel choice outcomes of children who watch cable.

Methods

Chapter 6 describes the methodology for this study in greater detail. Two-way cable data of household (not individual) viewing behavior are analyzed. Survey data from a subset of households wired for two-way cable measurement permitted identification of households with children under 18 living at home. Mail surveys were distributed first. Those who did not return a survey were called and interviewed by phone. One hundred fifty-three of the 200 sample households were reached and surveyed. Of those, 40 had children under 18. The sample size of households with error-free, two-way cable-viewing data varied from 110 total households on the worst day to 165 on the best days. Of the maximum 165 usable households, 33 were not surveyed, and they may or may not have included children under 18. If one assumes that the same proportion of households not surveyed have children at home as those which were surveyed, one would estimate that 7 of the 33 nonsurveyed homes had children at home. For the Saturday morning analyses two groups are compared: households which definitely included children under 18, and households which definitely did not or were among the 33 unknown. The decision to include nonsurveyed households in the nonchild sample was made in order to increase sample size.

Saturday mornings were selected as the viewing period most likely to include the largest child audience, and the smallest adult audience. Nielsen estimates indicate that children and teens comprise 71% of the audience on Saturday mornings, at least in noncable households (Banks, 1980). Data were collected for three Saturday mornings, 8 a.m. to noon, on the second, third, and fifth Saturdays of the 1982 fall season.

Again, the sample sizes are very small. The number of *viewing* households with children was 23, 27, and 27, respectively, across the three Saturdays. The number of other viewing households (called "other households" rather than "nonchild households" because up to seven may include children) was 57, 58, and 69. Even in households with children, there was no way of knowing whether a child was watching, although that likelihood is strongest on Saturday mornings. All data reported must be regarded as extremely tentative and suggestive.

The same variables used in Chapter 6 are also applied to this child analysis.

Results

Table 1 presents comparative data on viewing style. The households with children exhibited extensive channel changing. In the second and third weeks of the new season, at least one third of their viewing time was spent sampling channels, watching in strings of 14 or fewer consecutive minutes per channel. In Week 5, the level was lower, but still comprised one fourth of all viewing time. Among the other households, channel sampling (0–4 minutes), and particularly, extended sampling (5–14 minutes), was lower than in child households.

Comparing the Saturday morning viewing styles to viewing styles reported for prime time (Chapter 6, Table 4), both the child and other households watched more in short blocks of time on Saturday mornings. Maxi-stretch viewing (25 + minutes) comprised an average of 72% of prime time viewing time, compared to 64% of Saturday morning.

The share of viewing time to different types of channels (Table 2) reveals additional child versus other household differences. Network programming during the sample period (8 a.m.-noon) was comprised exclusively of shows targeted for children. Not surprisingly, the child households watched more network programming (61% of viewing time on the average), compared to other households

Table 1.   Saturday Morning Viewing Style Comparison: Percentage of Viewing Time

	Households with Children					
	10/2		**10/9**		**10/23**	
Sampling	11.3%		13.8%		8.8%	
(0–4 minutes)		36.4%		34.1%		25.1%
Extended Sampling	25.1		20.3		16.3	
(5–14 minutes)						
Stretch	5.5		11.2		7.1	
(15–24 minutes)						
Maxi-Stretch	58.0		54.7		67.5	
(25+ minutes)						
# viewing households	23		27		27	

	Other Households					
	10/2		**10/9**		**10/23**	
Sampling	10.7%		9.4%		8.0%	
(0–4 minutes)		27.5%		22.5%		19.8%
Mini-Stretch	16.8		13.1		11.8	
(5–14 minutes)						
Stretch	11.3		11.5		8.4	
(15–24 minutes)						
Maxi-Stretch	61.3		66.0		71.7	
(24+ minutes)						
# viewing households	57		58		69	

(36%). Other households watched more cable and non-network off-air channels. Channels available only over cable attracted one third of child household viewing time, suggesting that children do watch those channels, even when network programming is specifically designed to attract them.

More than half of the other household viewership was of cable-only channels, and another 11% of other off-air channels. Among nonchild households that did watch Saturday mornings, that period appears to be a prime time for cable and nonnetwork channels. Interestingly, nonchild households that watched television on Saturday mornings watched for as much time as child households.

The small sample sizes make interpretation of trends over time questionable. Considerable fluctuation in shares across different days can be seen (for example, 73% network share for child households on 10/2, down to 54% the next Saturday).

The shares of channel changes by type of channel follow the same pattern of variability (Table 3). Unlike prime time viewing, the network share of channel changes did not remain constant; it varied by as much as 20% among child households. For all types of channels, child households changed channels more times per hour. In fact, for all nonnetwork channels, they changed channels 10 or more times within every hour they watched. Other household channel changing

Table 2. Saturday Morning Channel Type Comparison: Percentage of Viewing Time

	Households with Children			
	10/2	10/9	10/23	Average
Networks	72.6%	54.4%	53.1%	61.4%
Other off-air	1.3	5.2	7.6	4.6
Pay cable	5.3	16.1	15.1	12.2
Other cable	20.6	24.1	24.0	22.0
total viewing time	1:52	1:51	2:10	1:58
# viewing households	23	27	27	

	Other Households			
	10/2	10/9	10/23	Average
Networks	43.3%	40.3%	22.1%	36.0%
Other off-air	7.9	12.4	15.7	11.4
Pay cable	17.5	18.3	13.8	16.5
Other cable	31.2	28.6	48.3	35.5
total viewing time	1:47	1:37	2:12	1:52
# viewing households	57	58	69	

Totals may not add to 100% due to rounding.

occurred less frequently, but followed the general pattern of least changing on network channels. The nonchild households exhibited decreasing channel changes per hour as the season progressed, from 7.2 changes per hour in the first week to 5.4 in the fifth week, paralleling the prime-time patterns. For the child

Table 3. Saturday Morning Channel Type Comparison: Percentage of Channel Changes

	Households with Children				
	10/2	10/9	10/23	Average	Changes/Hour
Networks	56.0%	35.6%	42.8%	44.8%	5.7
Other off-air	4.4	9.4	11.5	8.4	13.6
Pay cable	12.7	17.4	17.8	16.0	10.0
Other cable	26.7	37.4	27.7	30.6	10.2
# changes/hour	7.9	8.7	6.7	7.6	
viewing households	23	27	27		

	Other Households				
	10/2	10/9	10/23	Average	Changes/Hour
Networks	31.6%	30.6%	24.9%	29.0%	3.8
Other off-air	13.5	12.5	16.8	14.2	7.5
Pay cable	18.9	16.1	17.9	17.6	6.7
Other cable	36.0	40.5	40.4	39.0	6.8
# changes/hour	7.2	6.4	5.4	6.3	
viewing households	57	58	69		

households, that pattern didn't hold, although small sample size and instability may have been factors. The fifth week did find the fewest average channel changes.

Table 4 presents child and other household viewing shares for each individual channel and type of channel, averaged across the three sample Saturdays. Only six channels received a 6% or higher child household viewing share. Those receiving 2% or more included:

ABC (34%), NBC (17%), CBS (9%), MTV (7%), TMC (6%), HBO (6%), Weather (3%), ESPN (3%), CNN (3%), a local independent (2%), and Nickelodeon (2%).

Among other households, viewership was spread across more different channels. The same six channels, plus ESPN, received a 6% or more viewing share. Those receiving 2% or more of nonchild viewing share included:

CBS (13%), ABC (13%), NBC (10%), HBO (8%), TMC (8%), MTV (6%), CNN (5%), WTBS (5%), USA (3%), three local independents (2%–3%), two PBS stations (2%), and local and public access (2%).

For both child and other households, the networks, MTV, and pay movie channels were the major programming sources. The child households in this sample made little use of other cable channels, and even Nickelodeon was barely watched, at least on Saturday mornings.

Although nonchild viewing was distributed across a somewhat larger set of channels (17 channels receiving 2% or more viewing shares, compared to 11 channels for child households), this is more likely an artifact of a smaller sample of child households rather than a sign of more diverse channel preferences among other households. In fact, child households checked more different channels than other households (an average of 6.4 versus 5.6) and watched more different channels for 5 minutes or more than other households (3.2 versus 2.6).

CONCLUSIONS

On the basis of limited data reported here, it appears that the primary use of cable channels by the sample 23–27 children was watching pay cable and MTV. Beyond that, at least during Saturday mornings, few other cable channels were watched, and none for more than a 3% share of viewing time. Nickelodeon drew only 2%.

Why were so few cable channels watched? Are the children aware of what is available on most channels? To properly address these issues, a larger sample is needed—a sample which isolates viewing of different age groups.

As far as how children watch cable, these data suggest that children are active channel changers, often watching a channel for less than 15 minutes. Their chan-

Table 4.   Saturday Morning Individual Channel Viewership

Channel	Child Households		Other Households	
NETWORKS	61.4		36.0	
9 ABC		33.9*		12.5*
12 CBS		09.2*		12.8*
7 NBC		16.9*		09.8*
18 ABC		01.4		00.9
OTHER OFF-AIR	04.6		11.4	
16 PBS		00.1		01.7*
21 PBS		00.5		01.5*
19 Independent		02.2*		02.2*
22 Independent		00.1		00.9
28 Independent		00.5		02.5*
34 Independent		01.2		02.6*
PAY CABLE	12.2		16.5	
2 HBO		05.9*		08.4*
14 TMC		06.3*		08.1*
SUPERSTATIONS	02.1		06.8	
11 WOR		00.6		01.3
17 WTBS		00.9		04.5*
20 WGN		00.6		01.0
TEXT	04.2		04.0	
10 Program Guide		00.1		01.2
13 Weather Channel		03.3*		01.1
24 Cable Shopper		00.0		00.0
25 Newspaper		00.8		00.4
8 Stock Channel		00.0		01.4
OTHER CABLE	14.4		21.6	
15 USA		00.5		02.8*
23 CBN		00.0		00.3
26 MTV		06.7*		06.3*
27 ESPN		02.6*		05.8*
29 CNN		02.6*		04.6*
30 SNC		00.1		00.7
31 SPN		00.0		00.5
32 Nickelodeon		01.7*		00.6
33 PTL		00.1		00.0
35 C-Span		00.0		00.0
36 SIN		00.0		00.0
ACCESS	00.4		02.8	
3 government		00.0		00.0
4 local origination		00.3		01.7*
5 educational		00.1		01.4*
6 public access		00.0		00.0

*Indicates channel receiving a 2% higher of viewing share (rounded). Figures report share of viewing time to each channel and type of channel, and may not add to 100%, due to rounding. For a complete description of the channels available, see Appendix A.

nel sampling extended across the different channel types, even though only a limited set of channels attracted their attention enough to be viewed at length.

What inner processes does child channel changing represent? Is it a search for something to watch? How do they select which channel to watch? How do children deal with 35 or more channel options? Does TV viewing style parallel children's sophistication in decision-making skills in general? Is it developmental? Is viewing style modeled within a family? Will boys be found to change more than girls, paralleling adult findings of men changing more than women?

Or are the channel changes more often movement away from program content that doesn't hold a child's interest? How does channel changing relate to attention span? Research suggests a relationship between age and channel changing, with young viewers changing the most and old viewers changing the least. Why do children change more than adults, and younger adults change more than older adults? Will today's young, frequent changers grow up to be less-frequent changers, or is the current generational difference a function of some factor other than a developmental difference? For example, does early experience with television and remote-control devices result in more-active channel changing at all stages in life?

The major conclusion from this examination of child viewing behavior with cable is a realization that virtually nothing is known about what children do with cable. The area is ripe for systematic research, controlling for age and cable-system attributes and examining what children do with cable—how familiar they are with the different channels, their channel-use patterns, program-choice behavior, modeling of channel use and program choice, and attitudes toward cable, among other issues.

## CHAPTER 8

Changes in the Viewing Process Over Time

Bradley S. Greenberg

INTRODUCTION

The previous two chapters examined viewing style changes over time, across a 5-week period, using aggregate, two-way cable viewing data. This chapter takes a more macro approach to changes over time in the context of a 6-month field experiment. Individual behavior is assessed by survey, permitting analysis of changes over time and examination of the relationship of subscriber status (e.g., pay versus basic) to those changes. The issue of stability of viewing style is addressed. What kinds of stability are there in how television is watched over a 3-month or 6-month period? Do viewers develop a different style or approach in how they choose channels? Do they do more channel-flipping, or less? More zapping of commercials, or less? The luxury of examining such behavior over a year or longer would be even more ideal, but within what we have had available, a first step in looking at over-time changes in cableviewing is provided in this chapter.

So, what is expected? Begin with the three-segment process we have moved to, with phases of prior planning, orienting searches, and reevaluations. Over the course of a television season, from its beginning in the fall to its inevitable apex in the spring (at least for broadcast television), it seems reasonable to expect viewers to settle in among their preferred program/channel choices. A cyclical pattern of viewing style is proposed, which repeats each season. Early experimentation with alternative programs should lead to a winnowing; checking channels to see what's on should be less necessary; checking a guide, to the extent it was ever done, should produce less new information or information more accessible in standard on-air promotions. In other words, habit strength is expected to accumulate. In terms of prior planning, then, *viewers would be expected to make less use of guides and be more aware of what shows are on they want to watch.* They ought begin to watch essentially the same lineup of shows they watched last week and the week before that, and the same ones they expect to watch next week.

A similar logic would predict what goes on with their orienting searches. Whether they typically go in numerical order or some other order, whether they usually do a wide search before coming back to the best selection they could find, or whether they watch whatever happens to be on the channel when they turn the set on rather than checking around, the basic expectation is that *the orienting search will become somewhat truncated.* If they know what's on, or what it is they want to watch, or become creatures of week-after-week program schedules, they do not require the same magnitude of orientation in order to make a choice. They ought to open their television viewing by checking fewer channels.

Moving among programs or across commercials by changing channels before, during, and after program and commercial messages is anticipated to change little over time. The strength of this habit is less a function of time, we believe, than a purposive, deliberative maximizing behavior. This process of reevaluation is made so simple with remote-control devices (and our attention span may be so weak) that those who choose to get their exercise this way are likely to continue to do so. In fact, it is possible to argue that the avid channel changer derives no little pleasure in increasing his or her skill in flipping channels. However, to the extent that we have argued for more prior planning and reduced orienting when watching favorite shows and channels, *it is possible that channel changing during favorite shows* (not at commercials), already demonstrated to be a weak behavior, *would be diminished slightly. Channel changing during commercials and between shows probably continue unabated.*

At the same time, there are different groups of cableviewers by subscriber status, a factor more thoroughly examined in Chapter 16. Changes in cableviewing behaviors by subscriber status can also be considered. The essential differences among basic status, a single pay channel, and multiple pay channels are of both magnitude and kind. Obviously, you have more total channels if you pay for extra ones, and, equally obviously, you have different program content available, notably movies on the predominant pay channels. As for there being more channels, what kind of difference does it make if there are 36 available instead of 33 or 34? Likely none at all, and certainly not like moving from half a dozen off-air channels to a full cable system of 24-plus channels. Movies have no commercials to zap—they were designed for uninterrupted viewing. On the other hand, you are not confined to a single showing of a film; most of them, you can watch partly tonight and additionally on some other evening, if you wish—at no additional cost. So, we also examined what kinds of cable viewing differences occur across subscriber groups, and within each of them over time. Admittedly, this was an exploratory, descriptive examination without clear expectations.

METHODS

These data were gathered in conjunction with a field experiment conducted in cooperation with Continental Cablevision. The entire set of procedures is described in Chapter 20. These data were not a central part of the experiment and were gathered for the specific purposes reflected in this chapter.

Sample.

In brief, there were three waves of telephone interviews with two different samples. The first field survey was run during July 1984, the month prior to publishing the first issue of a new cable television guide. The second survey was completed 3 months later, after the third issue of the guide had been distributed. The final survey was run 6 months after the first survey, after the fifth issue of the guide had been published in December 1984.

Sampling details for the three waves are as follows:

1. For the first wave, a simple random sample of basic cable subscribers, HBO subscribers, and two-pay (HBO and Cinemax) subscribers was drawn from computer listings at Continental; in addition, a nonsubscriber sample was created by a random telephone-directory sampling. Interviews were completed with 209 basic, 104 one-pay, 109 two-pay, and 194 nonsubscribers. These represented completion rates of 72% for the cable groupings and 55% for the non-subscribers.

2. In Wave 2, computer-drawn samples of subscribers were created, excluding those used in the first wave. There were samples of 300 each for the three subscriber groups, and the completion rate was 72%, or 650 completed interviews. Only cable subscribers were surveyed.

3. In Wave 3, attempts were made to re-interview the group contacted at Wave 2; in all, 477 interviews were completed, or 73% of the 650 interviewed in Wave 2.

Variables.   An identical set of 14 questions was asked each time. These reflected a major portion of the change process variables as they had been operationalized to that point in time. For each question, the set of response categories was:

Very Often, Quite Often, Often, Not Very Often, Not at All (scored 1 . . . 5)

Two questions dealt with prior planning or *preselection*:

On weekday evenings, how often do you watch the same shows you watched the week before?
How often do you know what you are going to watch before you turn the TV set on?

Six questions dealt with the *orienting search* used by the viewers:

When you change channels to see what is on, how often do you stop changing channels *as soon as* you find something good?

When you change channels to see what is on, how often do you check many channels *and then go back* to the best one?

When you turn the TV set on, how often do you change channels to see what else is on?

When you check channels to see what's on, how often do you start with Channel 2 or 3, then go to 4, 5, 6, and so on in order?

When you check channels to see what's on, how often do you check them in some particular order *other than* 2, 3, 4, 5, 6, and so on?

When you check channels to see what's on, about how many different channels do you check *before you decide what to watch*? (for this question, an actual number was recorded for the response).

For the process of *reevaluation*, after a show has been chosen, four questions focused on channel-changing or zapping and two with other aspects of reevaluation. The channel changing questions were:

How often do you change channels or stop watching TV *before the show you are watching is over*?

How often do you change channels right *after* a show ends?

How often do you change channels *when commercials come on, during a show*?

How often do you change channels during a show, *not at commercials*?

Two other reevaluation questions were:

How often do you watch more than one show at a time by changing back and forth?

How often do you watch a whole show, once you've started to watch it?

Analyses.   From the first wave, we can re-examine these behaviors as they occur among subscribers and nonsubscribers, and then among the three different subscriber groups. Then, with the addition of the final two waves, in which attempts were made to re-interview the same respondents, we can look for change between the two time periods (across the entire sample). The final analyses will take us across all three time periods, for each of the subscriber groups—basic cable, one-pay, and multi-pay.

RESULTS

A first comparison between subscribers and nonsubscribers at a single point in time is examined in other chapters as well, particularly Chapter 16; it serves as a starting point for defining this sample and provides corroboration for the other findings. There are substantial initial differences between subscribers and non-subscribers in their viewing process—after the prior planning phase (see Table 1). Before starting to watch television, they behave the same, e.g., are equally likely to know what they're going to watch before they turn the set on, and equally likely to watch the same shows as the week before.

But the difference is fierce once they get in front of the television set. The main behavior at the time of initially choosing something to watch is that cable subscribers do considerably more channel checking; after all, they do have more available channels to check. Cableviewers check more than twice as many channels at this time, they are more likely to check them in some sequential order, to sample a lot of channels before going back to the best one, and to keep checking.

That behavior continues once a program has been chosen for viewing. Cable viewers do more channel changing at all time junctures—when ads come on, after the show ends, during the show itself, and before the show ends. And cable viewing is more likely to include attempts to watch more than one television show at a time, flipping back and forth and following two story lines or a ball game and a drama. Only the tendency to watch whole shows (not necessarily without changing around during those shows) is similar to the subscribers and nonsubscribers.

Table 2 takes the same variable sets and compares cable subscribers with basic-only subscriptions, a single-pay channel, and two-pay channels.

The overall trend found in the tabled data is for the pay subscriber to differ from the basic subscriber. Differences between single pay and multi-pay sub-

Table 1.   Wave 1: Process Variables by Subscribers and Nonsubscribers

	Type		
	Non-subs (n = 198)	Subs (n = 411)	p
Preselection			
a. Watch same show as week before	2.73	2.82	ns
b. Know before turning on TV	2.35	2.46	ns
Orienting search			
c. Stop changing at first good show	3.21	2.96	.02
d. Check many channels, go back to best one	3.46	3.02	.001
e. When turn TV on, change to see what else is on	3.60	3.32	.01
f. Check in order, 3, 4, 5, etc.	3.91	3.32	.001
g. Check in some other order	3.96	3.67	.001
h. How many do you check	3.16	6.79	.001
Reevaluation			
i. Change before show ends	3.79	3.66	.09
j. Change after show ends	3.47	3.01	.001
k. Change when ads come on	4.04	3.53	.001
l. Change during show, not at ads	4.34	4.07	.001
m. Watch >1 show at a time	4.38	3.82	.001
n. Watch whole show	2.20	2.20	ns

Table 2.   Wave 1: Process Variables by Subscriber Type

	Type				
	Basic **(n = 207)**	**1-Pay** **(n = 109)**	**2-Pay** **(n = 95)**	**p**	**p***
Preselection					
a. Watch same show as week before	2.70[a]	2.88	3.00[b]	ns	.06
b. Know before turning on TV	2.38	2.52	2.55	ns	
Orienting search					
c. Stop changing at first good show	3.03	2.82	2.97	ns	
d. Check many channels, go back to best one	3.05	3.11	2.84	ns	
e. When turn TV on, change to see what else is on	3.36	3.43[a]	3.13[b]	ns	.08
f. Check in order, 3, 4, 5, etc.	3.51[a]	3.10[b]	3.16[b]	.03	.02
g. Check in some other order	3.62	3.76	3.70	ns	
h. How many do you check	5.62[a]	7.43[b]	8.45[b]	.02	.01
Reevaluation					
i. Change before show ends	3.68	3.65	3.64	ns	
j. Change after show ends	3.08[a]	3.14[a]	2.72[b]	.02	.02
k. Change when ads come on	3.61	3.52	3.38	ns	
l. Change during show, not at ads	4.06	4.13	4.04	ns	
m. Watch >1 show at a time	3.92[a]	3.76	3.64[b]	ns	.08
n. Watch whole show	2.17	2.20	2.26	ns	

*The first p value is the Anova result. The second is from contrast tests and means with different superscripts are significantly different.

scribers are less pronounced and less consistent. Basic subscribers were more likely to follow a fixed television schedule, watching the same shows week after week. Pay subscribers were more active in their orienting searches than basic subscribers—they checked more total channels and they were more likely to check channels in numerical sequence when deciding what to watch. In terms of reevaluation, multipay subscribers were the most likely to change channels after the show they were watching ended, and to watch more than one show at the same time. Commercial zapping was similar across the three subscriber groups, as was channel changing before and during a show, other than at commercial breaks.

Changes Across Time

The data provide the opportunity to look at changes over time in how television is watched. Most directly, a comparison of the sets of respondents at Waves 2 and 3 was made. Recall that the Wave 3 interviews constituted a subset of the same respondents and households as at Wave 2, 3 months later.

Within this time frame, the obtained differences that are significant or border on significant fit into a meaningful pattern (see Table 3). As the season progresses, the cable viewers are:

Table 3.   Process Variables Between Waves 2 and 3

	Wave		
	2 **(n = 304)**	**3** **(n = 234)**	**p**
Preselection			
a. Watch same show as week before	2.45	2.30	ns
b. Know before turning on TV	2.44	2.22	.04
Orienting search			
c. Stop changing at first good show	3.19	3.08	ns
d. Check many channels, go back to best one	3.10	3.35	.03
e. When turn TV on, change to see what else is on	3.43	3.40	ns
f. Check in order, 3, 4, 5, etc.	3.56	3.59	ns
g. Check in some other order	3.58	3.85	.02
h. How many do you check	7.13	6.71	ns
Reevaluation			
i. Change before show ends	3.75	3.79	ns
j. Change after show ends	3.09	2.99	ns
k. Change when ads come on	3.49	3.66	ns
l. Change during show, not at ads	4.08	4.15	ns
m. Watch >1 show at a time	3.77	3.97	.07
n. Watch whole show	2.12	1.96	.08

- more likely to know what they want to watch before they turn on the television set.
- less likely to have to check or want to check many channels to find their show.
- less likely to check channels in numerical order, out of sequence.
- more likely to watch a whole show, and
- less likely to watch two shows at the same time.

In addition, there is no change of substance or magnitude on any of the channel-zapping measures. They are no more or less likely to change channels before or after the show ends, during commercials, or during the show. The channel-zapping behavior appears to be a quite stable one, over this time period at least.

Most channel changing still occurs after a show ends, a lesser amount at the onset of commercials, still less before the show ends, and the minimal amount of zapping during a show itself.

What appears to be the case—over time—is that cable viewers become more familiar with their major preferences and orient to those more readily. Stated somewhat differently, they may well do less exploring and become more settled into a smaller set of preferences.

Changes by subscriber group.   The changes just described are based largely on the same sets of respondents and households. One other important comparison is to look across all three time waves, and to do so separately for the three different subscriber groups. That is, to what extent are the overall changes reflected equivalently among basic, one-pay, and multi-pay groups? This is contained in Tables 4–6.

Let us begin with the basic subscribers (Table 4). Overall, the obtained differences are reflected in the Wave-1 vs. Wave-3 comparisons, with Wave-2 data typically falling in between these two boundaries. First, there is significantly more preselection by Wave 3, on both preselection measures, and the trend is linear. There is more watching of the same show as the week before and greater likelihood of knowing what is going to be watched before turning on the television set.

Second, there is a change in the orienting search process that is logical from the preselection differences. By Wave 2, there is less checking of many channels and then a return to the best one, and, by Wave 3, there is less checking of the available channels in some habitual order.

Third, there are no obtained differences in channel reevaluation behaviors. Zapping, channel flipping, and changing continue to occur at the same relative rates among basic subscribers, across all three time periods.

So, what were identified as changes over time between the overall samples at Waves 2 and 3 are parallel to the differences found for the basic subscribers

Table 4. Process Variables by Wave Among Basic Subscribers

	Wave				
	1 (n = 202)	2 (n = 117)	3 (n = 99)	p	p*
Preselection					
a. Watch same show as week before	2.70[a]	2.41[b]	2.22[b]	.01	.02
b. Know before turning on TV	2.38[a]	2.28	2.08[b]	ns	.05
Orienting search					
c. Stop changing at first good show	3.03[a]	3.32[b]	3.16	ns	.05
d. Check many channels, go back to best one	3.05[a]	3.35[b]	3.33[b]	.07	.04
e. When turn TV on, change to see what else is on	3.36	3.50	3.47	ns	
f. Check in order, 3, 4, 5, etc.	3.51	3.68	3.74	ns	
g. Check in some other order	3.62[a]	3.56[a]	3.93[b]	.06	.05
h. How many do you check	5.89	6.46	7.20	ns	
Reevaluation					
i. Change before show ends	3.68	3.71	3.74	ns	
j. Change after show ends	3.08	3.26	3.15	ns	
k. Change when ads come on	3.61	3.60	3.85	ns	
l. Change during show, not at ads	4.07	4.08	4.17	ns	
m. Watch >1 show at a time	3.93	3.91	4.14	ns	
n. Watch whole show	2.18	2.15	1.97	ns	

*In Tables 4–7, the first p value is the Anova result. The second is from contrast tests and means with different superscripts are significantly different.

Table 5. Process Variables by Wave Among One-Pay Subscribers

	Wave				
	1 (n = 110)	2 (n = 95)	3 (n = 77)	p	p*
Preselection					
a. Watch same show as week before	2.87[a]	2.36	2.26[b]	.01	.01
b. Know before turning on TV	2.52	2.37	2.32	ns	
Orienting search					
c. Stop changing at first good show	2.85[a]	3.28[b]	3.15[b]	.05	.02
d. Check many channels, go back to best one	3.12	3.01[a]	3.42[b]	.09	.04
e. When turn TV on, change to see what else is on	3.41	3.58[a]	3.80[b]	ns	.04
f. Check in order, 3, 4, 5, etc.	3.10[a]	3.67[b]	3.40	.03	.01
g. Check in some other order	3.76	3.64	3.70	ns	
h. How many do you check	7.59	8.93[a]	6.55[b]	ns	.08
Reevaluation					
i. Change before show ends	3.64	3.84	3.71	ns	
j. Change after show ends	3.16[a]	3.08	2.86[b]	ns	.09
k. Change when ads come on	3.50	3.41	3.53	ns	
l. Change during show, not at ads	4.13	4.22	4.03	ns	
m. Watch >1 show at a time	3.74	3.72	3.75	ns	
n. Watch whole show	2.21	1.97	2.01	ns	

across the three waves, lending additional credence to the former, often quite small, changes.

One-pay subscribers (Table 5) are more likely to watch the same show as the week before (preselection) by Wave 3. There is within this subscriber group a distinctive difference in orienting-search behavior across time—they do less in the way of orienting-search behaviors, although here the differences are found more often between the second and third waves than between other periods. The one-pay subscribers are less likely to check many channels and go back to the best one, they are less likely to stop changing at the first good show, they are less likely to check and see what else is on, and less likely to check channels in numerical order. There is also a nonlinear drift away from the total number of channels actually reported to be checked.

As for reevaluation after a show has been chosen, there are no consistent differences or changes across time, only the tendency for more changing to occur after a particular show ends than was reported in the first wave.

The analysis of the multipay subscribers continues and extends the patterns of changes/differences over time (Table 6). More changes occur in this group than in either the basic or the one-pay groups.

First, there is more preselection after Wave 1, more watching of the same show on a week-by-week basis, as found in the other subscriber groups. Second, there is less orienting search behavior—essentially amounting to less channel

Table 6.   Process Variables by Wave Among Multipay Subscribers

	Wave				
	1 (n = 93)	2 (n = 69)	3 (n = 46)	p	p*
Preselection					
a. Watch same show as week before	2.99[a]	2.51[b]	2.59[b]	.05	.03
b. Know before turning on TV	2.56	2.60	2.30	ns	
Orienting search					
c. Stop changing at first good show	2.96	2.90	2.98	ns	
d. Check many channels, go back to best one	2.83[a]	2.80[a]	3.37[b]	.05	.03
e. When turn TV on, change to see what else is on	3.11[a]	3.04[a]	3.60[b]	.05	.02
f. Check in order, 3, 4, 5, etc.	3.15[a]	3.17[a]	3.60[b]	ns	.07
g. Check in some other order	3.70[a]	3.37[b]	3.91[a]	.05	.02
h. How many do you check	8.52[a]	7.73[a]	5.53[b]	ns	.04
Reevaluation					
i. Change before show ends	3.64[a]	3.66[a]	4.00[b]	.08	.04
j. Change after show ends	2.74	2.71	2.83	ns	
k. Change when ads come on	3.38	3.34	3.57	ns	
l. Change during show, not at ads	4.03[a]	3.83[a]	4.30[b]	.03	.01
m. Watch >1 show at a time	3.61	3.48[a]	3.98[b]	ns	.05
n. Watch whole show	2.28[a]	2.17[a]	1.87[b]	.06	.02

checking, whether to see what else is on, or checking in numerical order, or checking many before returning to the best one, as well as the total number typically checked. With one exception, the change in this behavior occurs between Waves 2 and 3—i.e., Waves 1 and 2 are similar to each other; given the relatively smaller n's in this breakdown, the consistency of the differences is marked.

Third, and only within this subscriber group, there is a changing pattern of reevaluation, or channel zapping. By Wave 3, the multipay subscribers do less channel changing (or stop watching completely) before the show ends, less changing during the show, more watching of entire shows, and try less to watch more than one show at a time. Overall, the multipay subscribers are much less active in the viewing process than they were at the first (or second) wave of interviews.

Guide users.   Television viewing process variables were examined separately for those among the samples who said they had seen the experimental guide and those who said they had not seen it. Recall that all these households (at Waves 2 and 3) had received the multiple issues of the guide. Four separate analyses were conducted: (a) guide users between Waves 2 and 3, (b) nonguide users between Waves 2 and 3, (c) guide users vs. nonguide users at Wave 2, and (d) guide users vs. nonguide users at Wave 3. Suffice to report here that guide use was not a factor in the change-process variables. No consistent differences, other

Table 7.   Process Variables by Wave Among All Subscribers

	Wave				
	1 (n = 411)	2 (n = 293)	3 (n = 225)	p	p*
Preselection					
a. Watch same show as week before	2.82[a]	2.43[b]	2.32[b]	.01	.01
b. Know before turning on TV	2.46[a]	2.40	2.22[b]	.07	.03
Orienting search					
c. Stop changing at first good show	2.96[a]	3.17[b]	3.11	.07	.03
d. Check many channels, go back to best one	3.02[a]	3.10[a]	3.36[b]	.01	.03
e. When turn TV on, change to see what else is on	3.32	3.40	3.38	ns	
f. Check in order, 3, 4, 5, etc.	3.32[a]	3.54[b]	3.58[b]	.03	.03
g. Check in some other order	3.67[a]	3.53[a]	3.85[b]	.02	.01
h. How many do you check	6.95[a]	8.26[b]	7.85[b]	.01	.01
Reevaluation					
i. Change before show ends	3.66	3.74	3.77	ns	
j. Change after show ends	3.01	3.06	2.98	ns	
k. Change when ads come on	3.53	3.48	3.66	ns	
l. Change during show, not at ads	4.07	4.06	4.14	ns	
m. Watch >1 show at a time	3.82	3.75[a]	3.95[b]	ns	.07
n. Watch whole show	2.20[a]	2.11	1.96[b]	.03	.01

than could have occurred by chance, were found in any of these comparisons, and they are therefore not tabled in this chapter. The guide did not impact on the preselection of television shows, or in the process of orienting one to choosing a show, or in the process of reevaluation of show selections already made.

DISCUSSION

Changes over time in viewing-behavior styles do occur, although the time period used in this study was relatively short; stability in certain viewing styles or patterns also persists. What happens over longer periods of time is unknown but intriguing; e.g., to what extent is an adult viewing style learned in childhood, does the child emulate the viewing style of an older sibling, or of a parent, are some viewing style changes cyclical, attributable to seasonal variations in television fare? Given the general paucity of data gathered more than once from the same subsamples in media research, it is refreshing to be able to examine at least one data set in which time is a key factor.

What we believe is occurring is a continuing acclimation to a multichannel environment, in which the channel repertoire has been selected from among all that is available, to which the viewer is increasingly oriented, and from which channel/program selections are more typically being made—to turn the viewing process sequence into the kind of looping that overtime research might suggest. What is needed, of course, is a naive sample of broadcast television viewers. Then we could give cable to a random sample (some with, some without, a variety of pay channels), withhold it from a random sample, and then do a series of replicated interviews before they have cable, while it is fresh and novel, as it wears on, etc. That seemingly would be the ideal situation to better examine the notions proposed with this data set. Lacking that, we will struggle with approximations.

Most consistently prone to time effects (and what goes on during that time) appear to be preselection behaviors. Although the guide itself had no discernible impact on viewing-process behaviors, there were changes that supported the notion that viewers become more knowledgeable of what is on that they want to watch, and that their watching, or at least a sizeable chunk of it, becomes more routine. This would appear to be a content-related behavior, i.e., subject to what it is that is available in a given season, and perhaps subject as well to the promotional efforts of the telecasters in making on-air or on-cable information plentiful.

In contrast, channel changing during television viewing time—or reevaluating what is being watched—shows fewer changes or differences across time. In fact, here is where one might expect habits learned early to be little affected across the years of television viewing. This may be less a content-related variable, perhaps, and more a pure stylistic behavior. There are those who prefer to watch with the channel changer at the ready, taking advantage of lulls within shows, commercial breaks, between program periods, etc., to check around the dial. If so, such behavior would be less subject to time considerations and ought to be a more enduring characteristic, learned early in one's television life and not easily forsaken. One wonders how such an individual style is accommodated in a group viewing situation, or held in check by the perpetrator.

Motives for channel changing also have not yet been examined in any depth. Certain suppositions have been made; e.g., people are bored with a program, they want to avoid advertising. One can, however, conceive of a much broader range of motives involved. For example, there is little risk involved in channel switching; i.e., the likelihood of missing something vastly important may be perceived as a minimal risk, and there is the opportunity to switch back in part-seconds. So another motive may be that something better or equally good is on, that a higher reward for viewing exists on another channel at this same time. What may develop from this is the ability to watch (the best) of more than one show at a time, a behavior examined here but not found to be strongly used. Again, there is a strong possibility that this kind of behavior may be developmental, i.e., learned over time, or, better still, more likely to be exercised by young viewers more familiar with the new television technologies. If so, the next generation of viewers would be expected to be far more adept and casual in its channel-changing propensity. Another motive may be that channel changing is there to be done, a habit, nervous or otherwise. A substantial study of the motives involved in channel flipping is warranted.

What stands out in this set of findings are the substantial differences among viewing groups more so than the longitudinal differences. Those without cable are quite different than those with cable in how they watch television, as well as how much television they watch and of what it consists. There are of course background differences between subscribers and nonsubscribers as well that may account for a portion of those viewing behavior differences. However, within the cable subscribers, consistent differences reappear between basic and pay subscribers as well. Thus, it would appear that the multichannel environment is an operative factor on the one hand, and the content available on special channels (at an additional price) is also operative to induce particular styles of viewing. And, over time, particular facets of an individual's viewing are prone to change.

Viewer Type, Channel Type, and Viewing Styles

CHAPTER 9

Viewing Style Differences Between Radio and Television

Carrie Heeter and Ed Cohen

INTRODUCTION

While social scientists largely ignore radio listening because it is a "secondary activity," national ratings services regularly and intensively survey radio listening behaviors in markets across the country, to determine how many and what types of people are listening to each radio station. Despite this mass of systemized scrutiny, little is known about how radio listeners select the stations they listen to. Instead of examining the choice process, the research focus has been on outcomes of that choice process.

In this chapter, the model of television choice articulated earlier in the book will be adapted to radio listening, to begin to assess how well the model fits across different media choice behaviors. In particular, four elements of the choice process will be assessed for radio and compared to television viewing patterns: *planned listening* (knowing what station to listen to prior to turning on the set), *orienting search* (the process, if it occurs, of determining content available on one or many stations before committing to a particular station), *reevaluation* (reconsidering a station choice after it has been made—changing stations), and *channel repertoire* (the set of stations regularly listened to).

First, relevant differences between television viewing and radio listening should be considered and applied to choice-process constructs.

High- Versus Low-Involvement Activities

Radio may be described as a low-involvement medium, because listeners don't pay undivided attention to it. Television can also serve a "secondary activity" or background function, but it is more often primary. In-car radio listening may be expected to differ from at-home radio listening, in that the range of other activities which can co-occur while driving and listening is very limited. Thus, in-car radio use should more often be a primary activity than at-home use. More active radio consumption (e.g., more frequent and critical reevaluation of content and station switching) should accompany primary listening.

> H1: There will be more frequent reevaluation for in-car radio listening than at-home listening.

Pursuant to the logic that more frequent reevaluation will occur for primary activities, television viewing is probably more primary than in-car radio use.

> H2: There will be more frequent reevaluation for television viewing than in-car radio listening.

Number of Channel Options

Relevant to channel repertoire is the number of radio stations available, compared to television channels. Between local in-car and at-home radio, the station options should be synonymous, assuming equivalent AM–FM access. The comparison with television becomes complicated. In large metropolitan areas there are as many as 50 broadcast radio stations, compared to approximately 10 broadcast television channels. In smaller urban or rural areas there may be as few as 5 to 10 receivable radio stations, compared to 1 to 3 broadcast television channels. Cable television further complicates the equation, providing as many as 50 to 100 options in large cities, 36 to 50 in medium sized cities, and 6 to 12 in small communities. Thus, the ratio of available radio stations to available television stations varies extensively by area.

Within cable television systems carrying 36 channels, an average channel repertoire of 10 channels has been found across several studies. For broadcast television, the ratio of channels watched to channels available is much higher—approximately five of nine available channels. Factors other than sheet number of radio stations available may also influence channel repertoire.

Guide Availability

Regardless of how many channels are available, television is much more likely than radio to be accompanied by a weekly or daily guide detailing programming on most channels. Daily or weekly radio content availability across all stations is rarely listed in guides. Listener awareness of alternatives is therefore left even more up to chance or purposive exposure to a station.

Part of the reason for lack of composite radio guides is the difference in programming. While television carries programs usually ranging from 30 to 90 minutes long, produced prior to air time, the programming unit for radio is often as short as a piece of music (2 to 4 minutes), with the surrounding content and selection of music occurring during or shortly before air time. Less user planning should occur with radio than television.

> H3:   Radio listeners will know what they will listen to prior to turning on the set less often than they, as television viewers, know what they will watch ahead of time.

Channel Tuners

Another major difference in television versus radio is how stations are tuned in. Broadcast television tuners predominantly still have discrete dial positions for each channel. Cable television offers push-button tuners with fixed channel positions, or digital numeric keypads. For at-home radio listening, most often there is a continuous dial and the listener must carefully locate and tune in desired stations. In-car listening offers an alternative—four or five buttons which can be quickly preset to specific stations, allowing for easy jumping among a limited number of stations.

Preset buttons would allow for fast switching among a small number of stations. Cable's options essentially allow for random access of any channel—not quite as easy as preset buttons. Broadcast television dials force exposure to the channels in between. The continuous dial of at-home radio listening seems least conducive to channel switching, because it is so much more difficult to return to a station once you have left it.

> H4:   There will be more reevaluation for in-car than at-home radio listening.

In general, the lack of available radio guides, frequently fewer station options, and difficult (in the case of at-home) or limited (in the case of preset-car-button) search modes might result in smaller channel repertoires for radio than television.

> H5:   There will be smaller channel repertoires for radio than television.

In addition to these hypothesized viewing-listening differences, this study will examine how radio and television choice processes compare.

> Q1:   Does the television choice model fit for radio? Are the relationships among the construct consistent?

It may be that choice process resides as much with the individual as with the medium. Perhaps a frequent reevaluator of television is also a frequent reevaluator of radio.

> Q2: Does an individual's television choice process correlate with his or her radio choice process?

Television research has found that active television viewers with large channel repertoires who reevaluate frequently tend to be younger males. Does this hold true for radio as well?

> Q3: Are the same demographics that are associated with television behaviors also associated with radio?

METHODS

A telephone survey in a medium-sized Midwest community was conducted. The interviews were conducted by an undergraduate telecommunication class in audience research, supervised by the authors. The field work took place in late 1984.

The sample list of phone numbers was drawn by the interval skip method, with a random start point, from the metropolitan area telephone directory. Randomization of the last digit permitted inclusion of recently changed and unlisted numbers.

The questionnaire generally took 8 to 10 minutes to administer. A minimum age of 18 was the only screening criterion used. Four hundred twenty-one interviews were completed, for a completion rate of 45%. One reason for the low completion rate was fewer callbacks than might normally be used, due in part to different interviewers for each session rather than a continuous group throughout the course of the fieldwork.

The sample was 53% male. Average age was 38, with a range from 18 to 92. One-third had attended or graduated from high school, half had also attended or graduated from college, and an additional 18% had attended post-graduate programs. Twenty percent had an annual household income of less than $15,000, 21% earned $15–25,000, 20% earned $25–35,000, 14% earned $35–45,000, and 17% earned more than $45,000.

Radio listening habits and radio and television choice behaviors were assessed, and the models compared.

RESULTS

Descriptive Results

The first section of the questionnaire dealt with radio listening habits. As a measure of their radio *channel repertoire*, respondents were initially asked what stations they listened to regularly. The mean number of stations cited was 1.73 for this unaided recall question. The next question dealt with whether the person listened to AM only, FM only, or both. Nearly two-thirds of those questioned listened to FM exclusively, while 4% listen only to AM. Based on the bandwidths they listened to, respondents were read a list of local stations available in the appropriate frequencies, omitting any stations they had already indicated listening to. For each station, they were asked whether they ever listened to it. A total of 16 stations were available. Respondents identified an average of three stations listened to, in addition to the 1.73 derived from unaided recall, yielding an overall station set, or repertoire, of just under five stations.

Choice processes were assessed next, with separate questions for at-home and in-car radio listening. Table 1 compares the responses. On the average, respondents listened to the radio for 2.3 hours at home and 1.3 hours in a car on weekdays. *Planning* was operationalized by asking how often the respondent knew what station they would listen to before turning on the radio. A four-point scale was offered: 4 = Very Often, 3 = Often, 2 = Not Very Often and 1 = Not at All.

Sixty-three percent said they very often know what station they would listen to before turning on the radio at home. Another 21% answered often, for a mean of 3.43. In the car, knowing ahead of time dropped slightly to 54%, with 20% answering often, an average of 3.19.

Respondents were asked how many different stations they checked before deciding what to listen to, as a measure of *orienting search*. For at-home listening, the mean was 1.87, with a mode of zero. In the car, the mean was higher: 2.18, with a mode of two.

Next, the construct of *re-evaluation*, or zapping, was operationalized for radio. For both at-home and in-car listening, respondents were asked how often they change stations when commercials come on, when the announcer is talking,

Table 1. Home and Car Radio Use: A Comparison

	At Home	In Car
PLANNING		
How often do you know what station you will listen to before you turn on the radio?	3.43	3.19*
ORIENTING SEARCH		
How many different stations do you check before you decide what to listen to?	1.87	2.18*
REEVALUATION		
How often do you change stations when commercials come on?	1.73	2.17*
How often do you change when the announcer is talking?	1.63	2.00*
How often do you change during music?	1.64	1.79*
How often do you change during the news?	1.55	1.72*

*All comparisons on this page are paired t-tests for nonindependent groups. All are significant at p < .01. for all but orienting search, which uses actual respondent estimates, the response categories were 4 = VERY OFTEN, 3 = OFTEN, 2 = NOT VERY OFTEN, and 1 = NOT AT ALL.

during music, and during the news. Commercials are the content most often zapped by changing channels. Announcers are next most likely, followed by music and news. Reevaluation occurs significantly more often when listening to the radio in a car than at home, for each type of program content assessed.

The four re-evaluation items were factor analyzed and found to be unidimensional for both car and home listening, with all four variables loading highly (above .5) on the factor. Additive scales combining the four were created for car and home reevaluation.

When asked how often they experimented with their listening at home, a majority said not very often. The mean was 2.02. There was more experimentation with listening in the car. The modal response was still not very often (48%), but the mean rose to 2.23.

Among the 78% who have a push-button car radio, an average of four and a mode of five buttons have been set to specific stations. Nearly three-fourths had not changed the stations the buttons were set for in at least a month. This suggests a stable in-car channel repertoire.

Media Comparisons

The radio choice items can be contrasted with similar questions about television choice. (See Table 2.) Planning for television viewing was significantly higher than planning for in-car radio listening, but not different from at-home radio use.

Respondents report a larger orienting search for television. TV viewers check an average of 3.76 different TV channels, compared to 2.18 car radio and 1.56 home radio stations when deciding what content to select.

Reevaluation of program content occurs most often for television, followed by car, and finally home, radio. Channel repertoire based on unaided recall measures is considerably larger (4.08 stations) for television than for radio (1.73 stations). Repertoire was not asked separately for car and home radio listening.

In Table 3, correlations between respondents' choice behaviors are compared across media. Car and home radio listening styles are highly correlated, ranging from .41 to .59. Planning for radio listening is unrelated to planning for televi-

Table 2. Radio–TV Mean Comparisons of Choice Behaviors

Variable	Car-Radio	TV	Home Radio
Planning[1]	3.19	*3.31	3.43
Orienting Search[2]	2.18	*3.76*	1.56
Reevaluation[1]	1.98	*2.21*	1.70
Channel Repertoire[3]	1.73	*4.08	

*Significant differences between television and the radio mean behaviors appear adjacent to the asterisk

[1] these items reflect a scale from 4 = Very Often to 1 = Not Often At all
[2] these items represent actual numeric estimates by respondents
[3] channel repertoire was not assessed separately for radio and car listening

sion viewing. For orienting searches, and particularly for re-evaluation, the tendency to zap in one medium is significantly related to a tendency to zap in the other. The size of respondents' channel repertoires is not correlated between radio and TV.

The choice model diagrams with correlations between each choice behavior variable are shown in Figure 1 for home and car radio and for television. Comparing across the three models, the correlations are remarkably similar. In all cases, the significance or lack thereof, and the direction of each correlation, is matched across the models. Thus, planning is always negatively related to an extensive orienting search and to zapping (re-evaluation). A larger orienting search and more zapping are related to watching or listening to more different channels. Orienting searches and re-evaluation tend to co-occur. And planning is not related to the size of the channel repertoire.

The major media differences are these: With television, the negative relationship between planning and orienting search is stronger, as are the positive associations of orienting search and re-evaluation with channel repertoire. With radio, there is a stronger relationship between orienting search and re-evaluation.

Table 4 shows the relationships between demographics and the media-choice behaviors. Age very consistently is related to choice process, across all media. The trend is for older adults to plan more, search and re-evaluate less, and listen to or watch fewer channels. The relationship of sex and choice behavior is weaker. With television, males typically zap more and engage in more extensive orienting searches. With radio, no difference appears for car listening, and males change slightly more when listening at home. With car listening, females are slightly less likely to plan their listening. Income is more related to radio listening style than TV viewing style. The higher someone's income, the more likely he or she is to plan listening and not to zap or use orienting searches. Education is only slightly associated with a few choice behaviors.

DISCUSSION

From this study, we can begin to describe radio zapping behavior. Not surprisingly, switching stations occurs more frequently when listening to radio while driving, with the radio in easy reach, than at home. More than three-fourths of

Table 3.   Intermedia Correlations for Choice Behaviors

Variable	Car-Home	Home-TV	Car-TV
Planning	.43*	.00	.00
Orienting Search	.41*	.20*	.11*
Reevaluation	.59*	.24*	.29*
Channel Repertoire[1]		.01	

*Indicates a significant correlation at $p < .05$
[1]radio repertoire was not assessed separately for home and car use

Figure 1. Radio and TV Choice Models

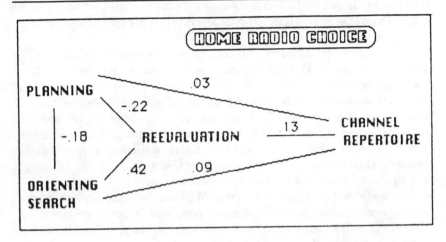

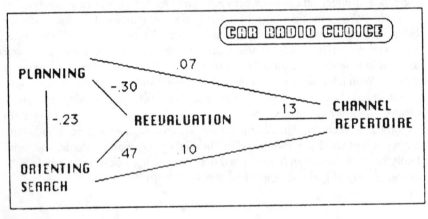

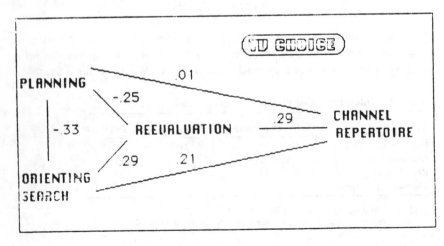

Table 4. Selected Demographics and Choice Styles

	Age	Sex	Income	Education
PLANNING				
car	.09*	−.09*	.15*	.08
home	.02	−.08	.10*	.16*
tv	.16*	−.05	.08	.02
SEARCH				
car	−.24*	.01	−.14*	.03
home	−.24*	−.09	−.11*	−.10*
tv	−.19*	−.12*	.01	−.06
REEVALUATION				
car	−.43*	.05	−.19*	.04
home	−.35*	−.09*	−.20*	−.03
tv	−.31*	−.19*	.04	.01
CHANNEL REPERTOIRE				
radio	−.13*	−.06	.01	.06
tv	−.07	−.06	.09*	.00

*indicates a significant correlation of $p < .05$.

car radio listeners have push-buttons which can be preset. They preset an average of four buttons, and change those settings infrequently. This survey found that 75% had changed settings no more recently than 1 month earlier, but it does nothing to identify how much longer ago the settings were actually changed. Even though many car listeners have four or five preset stations accessible at the push of a button, they check far fewer stations (1.86) before deciding what to listen to, and they often know what they will listen to before they turn on the radio. On the average, respondents listened regularly to 1.73 different radio stations, and occasionally to another three stations.

As far as *when* listeners zap, commercials are the most frequent content that gets zapped. From the questions in this survey, it is difficult to estimate how many commercials get zapped. The average response falls between "often" and "not very often," closer to "not very often." Announcers are second most likely to be zapped, and music and news least likely. Further research should seek clarification; e.g., in a half hour of radio listening, about how many times would you say you change stations? Self-reported estimates of such behaviors are not necessarily good indicators, and actual behavioral measures would be even more appropriate.

In general, television viewing is a more active pastime than radio listening. As viewers, these respondents reported that they planned more, checked more different channels, zapped more, and watched more different TV channels than they did with radio. And as already mentioned, car listening is more active than home listening. One explanation for this is that radio listening is often cited as a secondary activity, while TV viewing is a primary one. Similarly, car listening may be more primary than home listening, where other activities and distractions are more available. But that would not account for the difference in number of channels checked and watched between the media. Why, if there are five buttons preset to stations, does someone searching for something to listen to only check a few stations? Actually, the frequency distributions for these variables suggest that, as with television, there is a small subset of radio listeners who approach the label of "hyperactive" listeners, changing very often at commercials and announcers and during music and news, checking nine different channels, and so on. But overall, for the majority of home and car listeners, most listening is fairly passive.

In-car and at-home radio listening styles are highly related, and the frequent-channel-changing TV viewer tends also to change radio stations more often. The frequent changer tends to be younger and earn less income, although education does not make a difference. While TV zapping is more often a male phenomenon, radio zapping is not. Car radio zapping is not different by sex, and home radio zapping is significantly related to being female, although the relationship is not strong. Perhaps in the home, the female exercises more control over that medium, or listens and cares more what is on.

This study also further validates the choice model developed for television viewing style. The relationships between the choice variables for car radio are consistent with those for home radio and for TV to the extent that all correlations maintain the same direction, and, in each instance where a correlation in one model achieves significance, it does so in the other two.

There are two directions which this line of choice model research should now turn to. First, there has yet to be developed a strongly predictive description of viewing-style typologies. The motivation behind the zapping or planning or searching—be it a short attention span, need for variety, high IQ (or low IQ), etc.—has not really been identified. Secondly, the implications of the variations in viewing style have yet to be explored. What does it mean to have a large channel repertoire? How do channel repertoires develop? How do they change? Is the hyperactive listener/viewer paying more attention to what he or she consumes?

## CHAPTER 10

Viewing Context and Style with Electronic Assessment of Viewing Behavior

Bradley S. Greenberg, Carrie Heeter, and Sherri Sipes

INTRODUCTION

In Chapter 4, viewing behaviors from the vantage point of electronic assessment were characterized. That replaced the problematic self-reports of behaviors that respondents may not have given much thought to until they were asked to indicate what they did. Thus, the reconstruction of how often they did this or that—although demonstrated elsewhere in these pages to be reliable—still begs the validity question: Did they really do those things, some of them, all of them, and to what extent are they aware of how often?

Of course, electronic assessment of cableviewing behavior is no panacea. Chapter 4 detailed the extent to which breakdowns in the measurement system cut into the number of homes from which data were collected, the number of days for which it was available, etc. In addition, this examination of viewing-process choice making has focused almost exclusively on the individual and his or her behaviors. The electronic measurement process was *not* about individuals, but about households. Thus it corresponded to the typical form of obtaining television ratings electronically, e.g., Nielsen ratings. In recent years, however, renewed competition among ratings companies and demands for better data by the networks and the advertisers have lead to current efforts at using "people meters" to gain individual viewing data. However, the measurement system available in the Temple Terrace, Florida, setting was geared to assessing channel changes in far smaller time units than rating services provide. Nonetheless, it provided household information and did not tell us about individuals.

In this chapter, the Temple Terrace data are once again brought forward, and placed in the context of individual members of those households. The households that had been electronically assessed were returned to, and individual interviews were conducted with one adult member in many of those households. In this manner, we obtained a first estimate of linkages between viewer characteristics, principally media and self-reported viewing-style characteristics, with an independent measure of household viewing behavior, albeit not a strict measure of idiosyncratic viewing behavior.

Conceptually, then, this is what was examined:

The *demography* of viewers. . . .
    . . . as linked to their *viewing environment* or context
    . . . as related to their self-reported *viewing style* characteristic patterns
    . . . as predictive of their measured household *viewing behaviors.*

The choice of variables was, at the time of this study, largely a function of what prior literature suggested would be predictive of extended cable viewing and use, and to some extent a function of what was at this time just beginning to emerge for us as potentially significant viewing-process variables. As with many other studies in this volume, the research process was largely exploratory; nevertheless, specific expectations guided the selection of potential covariates of viewing.

Beginning with a small block of demographic characteristics that have previously been shown to be indicative of either cable subscribership or amount of cable-viewing behavior—age, gender, education, income, and number of adults and number of children in the home—the general expectation was that these would relate in similar ways to viewing style, both by the electronic assessment and self-report. If any demographics were to stand out as predictors of measured household viewing behavior, it would likely be that which tapped household size, whether in the form of adults or nonadults or both.

The notion of environmental variables included those which would both enhance amount of time spent viewing and mitigate against it. However, since a chief consideration of cableviewing in our conception of it was channel-changing behavior and decision making about what to watch, other considerations entered the choice of environmental variables examined.

Specific facilities available should enhance viewing, and these were considered to be the extent to which multiple sets in the household were serviced by the cable system, rather than a single set; similarly, the choice of adding pay channels to the basic service increased the channel options available to the viewer. Finally, as determined in other studies reported in this volume, the avid cable fan is more generally an avid technology fan, and it was anticipated that heavier viewing would be found among those with more "gadgets" in their home, e.g., VCR, computer, etc.

Viewing-context variables that were expected to impinge on the viewing process included the extent to which viewing was done alone, rather than in the company of others. One outcome of this viewing context might be more channel changing. We also sought to determine if more channel changing was sympto-

matic of busier viewers in general. Thus, a measure of alternative activities going on while watching television was developed, e.g., eating while watching,
playing games, etc. Whether this would identify a busier viewer, or a less atten-
tive viewer, or a more active, frenetic viewer was not well developed, but, given
much speculation and little concrete evidence of the kinds of alternative activities
people engage in while watching television, this seemed an appropriate opportunity to explore that aspect. Viewers were also asked to estimate their typical
weekday viewing in hours; having this to compare with electronic measures of
household viewing would provide a first step in knowing the fit between these
kinds of data.

Activities totally competitive with television viewing were also considered a
portion of the viewing environment (or perhaps nonviewing context). If an individual spent more time in social activities, or sporting activities, or other forms
of entertainment, there should be less time available for television. Similarly, if
the individual were an avid reader of magazines or newspapers, there should be
less time available for television; and perhaps the reading fan does not zap
around the channels, but is more informed and selective in his or her initial program choice.

The viewing-style or choice-process variables assessed in this study were a
subset of those reported on elsewhere in this volume, although here they have
some different properties. We were, after all, thinking ahead to the kinds of style
characteristics which might, from self-reports, be associated with the electronic
measures available. Those measures were largely constrained to time considerations (how much time was spent watching a given channel during a given time
period) and to channel-changing considerations (how often was a channel
changed, which channel was it).

In fairly straightforward correspondence to those eventual criterion variables,
then, respondents were asked to identify their favorite television channels and to
identify the channels they typically checked before they decided what to watch.
The magnitude of each of these was anticipated to be related to the actual
channel-changing behaviors obtained electronically. In similar fashion, if they
were prepared to watch television before entering the system, i.e., knew what
they wanted to watch, there should be less flipping, and so they were queried
about their use of television guides before turning on television.

In addition, two archetype viewing styles, or typologies, seemed worthy of
consideration at this point. These might be complementary or competitive. For
one, there is the viewer who would be considered largely nonselective; this
viewer would turn the set on, flip channels to find something that looks good and
stop there, or would turn the set on and indeed watch whatever channel it happened to be turned to at the time. For another, a more adventuresome viewer,
television watching would likely consist of trying to watch more than one show
at a time in order to maximize the intensity of the viewing experience in any
given time unit. Unfortunately, in retrospect, we did not create variables that
would have provided us with an assessment of "planned viewing" in which the

choices are preknown, not modified, and options while viewing are little explored; we also did not include what later became a standard set of questions in several other studies reported in this volume that tapped channel changing before and during shows, and during commercials.

The final variable in this set asked respondents how often there were disagreements over the channel-choice decision-making process in the household, and how often someone emerged upset with the outcome of this decision. It was anticipated that households with more frequent channel changing orientations would be households of greater perceived disagreement.

The set of household measures electronically obtained will be elaborated in the methods section. They constituted the dependent or criterion variables in this study and were various indices of viewing time, frequency of channel changes, and the number of different channels used in this 36-channel system.

METHODS

All households in the Temple Terrace area that had been wired for electronic measurement of their cable viewing behavior (n = 197) were mailed a questionnaire for completion by one adult individual in the household. Questionnaires were completed by 164 households; two-thirds of the completed interviews were obtained by return mail; for those not returned, a telephone interview provided the additional completions. From these completed households, there was reliable electronic data in 127, and that is the database used in this chapter. The electronic data span 3 June weekdays of viewing television.

In accord with the model developed in the introduction of this chapter, four sets of variables were obtained from this interviewing process, and one set of variables was obtained from the electronic measures obtained from the household. The four sets obtained by questionnaire included information about *demographic* characteristics, a set of measures more indirectly related to viewing (largely measures of the *viewing environment*), and direct viewing *process* and *outcome* behaviors. Those obtained from *electronic measurement of household viewing* will be described below as well.

Demography. This was a fairly affluent set of cabled households, not a retirement center in Florida; 29% reported household incomes greater than $50,000, 30% between $30–50,000, 26% from $20–30,000, and 16% below $20,000. The average age of the respondent was 47, and exactly half the respondents were females. The mean education was 2.5 years of college. Ethnicity was not used in analysis because the sample was 93% Anglo.

Within this set of demographic characteristics, significant relationships were found between age and number of children (− .44), and income with number of adults (.30), and with education (.19), gender (females) with income (− .20), and age (− .17).

Viewing-environment variables.   This set of measures examined certain aspects of the context of viewing, including competitive options. First, three ownership measures were assessed. An index labeled *gadgets* was created, summing ownership of other pieces of entertainment equipment: ownership of VCRs (20%), videodisc players (4%), home computers (15%), video games (28%), and video cameras (3%). *Cable* was examined in terms of the number of sets connected to the cable service (43% had more than one set connected), and *pay cable* was assessed in terms of the number of pay channels available in the home (11% had no pay channels, 22% had one pay channel, and 67% had both pay channels available in this system).

Next, three aspects of viewing were assessed. Respondents estimated the average weekday *TV hours* they watched (17% said 2 hours, 19% said 3, 17%–4, 15%–5, 11%–6, and 16%–7 or more). The frequency of watching television *alone* yielded another rectangular distribution: Given a five-position response scale anchored by "very often" and "almost never," 24% very often watched alone and 15% almost never did so, with intermediate proportions of 16%, 26%, and 19%. The third measure was an index summing across eight different *competing activities* that the respondents indicated they or other members of the household did while watching television—eating meals, eating snacks, reading, talking with others, playing games, working (or homework), housework, and craftwork. Those activities with the largest "very often" response were eating meals and eating snacks (both more than 30%); those with the largest "almost never" response were playing games, craftwork, reading, and homework (all over 40%).

Household or individual activities other than during television-watching time can also impact on the television-viewing process. Respondents were given 10 such activities to consider. Initial analyses of the relationships among these activities led to the creation of three leisure indexes (for each, the proportion indicating this was done "very often" is in parenthesis): the *sports* leisure time index consisted of the frequency with which respondents reported participating in sports activities (19%), playing indoor or outdoor games (20%), and attending sporting events (16%); the *social* index included how often they reported going to plays or concerts (6%), visiting at friends' homes (13%), going out to movies (4%), and going out dancing or drinking (6%); the *audio* index including listening to music (46%) and listening to radio news (28%).

The final measure in this set was use of *periodicals*—the number of newspapers and magazines respondents received regularly. On average, the sample received 1.5 newspapers and 3.5 magazines regularly.

Within this set of variables, the primary significant relationships were number of TV hours viewed—with gadgets (−.18), viewing alone (.27), sports activities (−.25), and social activities (−.21). In addition, leisure time given to sports activities was positively related to both social (.34) and audio (.24) activities, and (surprisingly) to the number of pay channels available (.18). Use of periodicals and number of cabled sets were unrelated to any other of these viewing-environment variables.

Process variables. In accord with the overall thrust of research in this volume, a series of questions asked how the viewer went about choosing what to watch on television. Several questions used the response format ranging from "very often" to "almost never." Respondents were asked how often they *checked a program guide before* turning on the set (54% did so "very often" or "often"). An index of *nonselective* choice summed the frequency with which they changed channels and stopped at the first show that looked good (34%) with the frequency of their watching whatever happened to be on (13%). *Adventurous* choice making was *not* knowing what they would watch before turning the set on (14%), in conjunction with watching more than one show at a time by changing back and forth between channels (47% do so). In addition, respondents were asked which channels they usually checked to see what was on, and the *number of different channels checked* by them was used (mean = 5, mode = 9).

Finally, two process-outcome variables were created. One asked respondents to identify their *favorite TV channels*, and the sum of those cited was used (mean = 3.5, mode = 3). The second determined whether the choice process used in the household produced disagreements (*upset*) among family members; i.e., how often there was disagreement about what to watch (53% "almost never"), and how often someone was upset by the outcome (64% "almost never").

Within this subset of process variables, nonselective viewing was related to finding someone upset with the choice process (.32) and with not checking a guide before viewing (. − 39); adventurous viewing was correlated with the number of channels checked (.17) and with checking a guide before viewing (.31).

Measured variables. Ten electronic measures were computed initially from each household:

1. the number of different channels viewed for 1 minute or less
2. the number of different channels viewed for more than 1 minute but less than 5 minutes
3. the amount of time viewed in units of less than 5 minutes (string viewing)
4. the number of channel changes made while viewing in strings
5. the amount of time viewed in units of at least 5 but less than 15 minutes (mini-stretches)
6. the number of channel changes made while viewing in 5- to 15-minute segments
7. the amount of time viewing in units of at least 15 but less than 26 minutes (stretches)
8. the number of channel changes made while viewing in 15- to 26-minute segments
9. the amount of time viewing in units of at least 26 minutes (maxi-stretches)
10. The number of channel changes made while viewing in 26 + minute segments.

Based on examination of the relationships among these measures, the following indexes were constructed.

  I.   Total television viewing time
  II. The total number of channel changes across all time segments
  III. Total viewing time in strings (<5 minutes)
  IV. Total viewing time in mini-stretches of 5 to 14-minutes
  V.   Total viewing time in stretches of 15 to 25 minutes
  VI. Total viewing time in maxi-stretches >25 minutes
  VII. Total channel changes in maxi-stretches

Even with this reduced set, there remain empirical redundancies among these measures, as evidenced in Table 1. Total viewing time is largely redundant with maxi-stretch viewing time and with channel changes in those viewing periods. Total channel changes is redundant with the amount of viewing in strings and in mini-stretches; the latter two are also very highly correlated. Stretch viewing time is substantially correlated (.5-.7) with all variables except maxi-stretch time.

RESULTS

Examining the results will begin by moving in sequence from the relationships between

> Demographic and Environmental variables
> Demographic and Process variables
> Environmental and Process variables

and following these with an examination of the separate and combined influences of these three sets of variables on the electronically measured set of variables, using multiple regression analysis at that stage of our model testing. Relationships reported throughout this section are $p < .05$, unless otherwise indicated; complete correlation matrices are available from the authors.

Table 1.   Correlations Among Electronic Measures

	Total Time	Channel Changes	String Time	Mini Time	Stretch Time	Maxi Time	Maxi Changes
Total Time		.44	.37	.46	.53	.92	.82
Channel Changes			.98	.88	.62	.08	.32
String Time				.86	.54	.02	.24
Mini Time					.68	.09	.34
Stretch Time						.22	.46
Maxi Time							.76

Demography and Environment

Media facilities—gadgets, pay cable channels, and cabled sets—were all positively related to income. The number of children under 18 also was linked to having more media gadgets in the home, whereas the number of adults was related to the number of cabled sets. Older respondents were more likely to have more cabled sets and less likely to have pay-cable channels.

Viewing alone was unrelated to any demographic characteristic in this set. Estimated hours of television viewing for the respondent was positively related to age and negatively related to education, income, and the number of children under 18. Doing other things while watching television was done more by younger persons and by females.

Among the alternative leisure-time activities, listening to music and radio were not related to the demographic traits; social leisure time was strongly related to age ($-.44$). The extent of participation in sporting activities was negatively related to age, but positively related to education, income, and the numbers of children and adults in the home.

The demographic characteristic with the most persistent set of relationships among these viewing environment variables is age.

Demographic and Process Variables

The closing sentence in the previous section should well be the opening sentence in this one, only more so. Both nonselective viewing and adventurous viewing are negatively related to age, whereas the number of children under 18 is positively associated with nonselective viewing and the number of adults is positively related to adventurous viewing.

Age is positively related to checking a television guide before turning on the set and the number of favorite television channels; it is negatively associated with changing channels to see what's on and with perceiving that anyone is upset during the viewing choice process. As expected, the more children in the house, the more disagreement over choices.

Environment and Process

Media facilities in the home are little related to the viewing-process variables. There is more disagreement in an environment with more gadgets and more pay-cable channels, and more channels are checked if you have more pay channels, but these are the only significant relationships with the process behaviors.

Viewing alone is related only to nonselective viewing (watching whatever happens to be on), and total estimated viewing hours is related only to the number of favorite channels identified by the respondents. However, doing other things while watching television is more generally linked to viewing-process behaviors: it is positively related to nonselective viewing, to checking more channels when trying to decide what to watch, and to the feeling that people are upset by the choice process. Doing other things while watching is negatively related to checking a television guide before turning on the set.

Social activities during leisure time also show a broad range of significant relationships with the process variables. The more social activities, the more nonselective *and* adventurous viewing, the fewer favorite channels, and the more disagreement found in the viewing choice process. Disagreement also co-occurs among sporting enthusiasts. Avid readers do less nonselective viewing and have more favorite channels.

Model Building

Now the effort turns to using these self-report pieces of information to gauge true behavior, as assessed by electronic measurement of television set use. The strong caveat remains that the electronic assessment was of household behavior, rather than individual respondent behavior, a shortcoming that current ''people meter'' research is able to rectify if it chooses to make measurements parallel to those reported here.

To better understand the relationships between the several sets of variables presented and the electronic measures, stepwise multiple regression analyses have been conducted—first separately for demographic, environment, and process variables, and then with all these sets pooled.

Television viewing time.   Regression results are in Table 2. Two of six demographic attributes provide significant relationships with the total viewing time in the household—the more adults (18 +) there are living in the household, and the lower the educational level of the respondent, the more time the television set was turned on. Two home-environment variables also provide a significant multiple correlation; the more communication gadgets in the home, e.g., VCR, personal computer, the more television is watched. In addition, the individual's estimate of his or her television viewing time on an average day is a significant correlate of how much time the set actually is on. Further, the tabled data include those variables which met a lesser criterion of statistical significance; the amount of sports activities engaged in by household members (as reported by the respondent) tended to contribute to TV-viewing time. No viewing-process variables were significantly related to total viewing time.

The overall solution, which tests all variables at the same time against the criterion variable (i.e., total viewing time in this analysis), retains three of the four individual variables just cited (number of adults, number of gadgets, and respondent's education). In addition, leisure time spent in audio activities, i.e., listening to music and to radio news broadcasts, tended to be part of the explanation of viewing time. Given the final multiple R = .468, the three remaining variables accounted for 22% of the variance associated with total household viewing time.

**Table 2.   Multiple Regressions Predicting
Total Television Viewing Time**

	R[1]	Beta
Demography		
# of Adults in Home	.262	.250
Education of Respondent	.332	− .204
Environment		
Gadgets in Home	.179	.214
Estimated Daily TV Hours	.261	.193
(Sports Leisure Activities)		(.179)*
Process	(*no significant correlates*)	
Overall Solution		
# of Adults in Home	.371	.341
Gadgets	.421	.241
Education of Respondent	.468	− .207
(Audio Leisure Activities)		(− .172)**

[1]The multiple R is cumulative across the variables within
each subset; the Beta is that obtained from the solution which
includes all significant variables listed, given a criterion of .05
for inclusion. Parenthetical variables were not included in the
solution because they did not meet the significance criterion, but
are presented here for their informational value.
     *p = .051; **p = .072

Total channel changes.   Table 3 presents the results. Those significant correlates from among the demographic characteristics were the number of adults in
the household and the age of the respondent; the more adults and the younger the
respondent, the more channel-changing behavior in any unit of viewing time.
Environmental correlates of this measured behavior were the amount of communication gadgets found in the home, time spent with sporting activities, time
spent with listening activities, and, to a lesser extent, the amount of television
watched with no one else present. Among these, all had positive relationships
with total channel changing, except for listening activities. The more listening
activities were engaged in, the less channel changing occurred.

   For this behavior, process variables also provided a significant multiple correlation, comprised of adventurous viewing and the extent to which people in the
household were upset or disagreed over the viewing choices made. For both
these, positive relationships were obtained—the more adventurous the viewer
(e.g., watching two shows concurrently), and the more disagreement over
viewing choices, the more channel changing was found.

   However, in the overall multiple regression solution, the process variables
dropped out. What they were measuring was redundant and better accounted for
by four correlates of channel changing—number of adults, leisure time given to
sports and listening activities, and the number of newspapers and magazines
available in the household. Adult frequency and sports activities were positively
correlated with channel changing; audio activities and the use of periodicals were
negatively correlated with channel changing.

**Table 3. Multiple Regressions Predicting
Total Channel Changes**

	R	Beta
Demography		
# of Adults in Household	.217	.223
Age of Respondent	.306	−.216
Environment		
Gadgets in Home	.332	.286
Audio Leisure Activities	.390	−.234
Sports Leisure Activities	.445	.218
(Watching TV Alone)		(.154)*
Process		
Adventurous Viewing	.372	.354
Upset from Choice Process	.412	.178
Overall Solution		
Adults	.264	.247
Audio Leisure Activities	.380	−.316
Sports Leisure Activities	.447	.284
Use of Periodicals	.498	−.224

*p = .073

These four predictors accounted for 25% of the variance in the behavior of channel changing.

Time viewing in strings.   Short viewing spurts—defined as viewing in segments of less than 5 minutes—was the next criterion examined (Table 4). Age of respondent and number of adults were again the only significant demographic correlates. The environmental variables were the same subset that predicted channel changing—communication gadgets in the home, sports and audio leisure time, and watching alone. Given the complete redundancy between channel changing and time spent viewing in strings ($r = .98$), comparability in the regression results is mandatory. Among the process variables, adventurous viewing provided the only significant relationship. Then, in the overall solution, age and gadgets dropped out, as did adventurous viewing, leaving adults, audio and sports activities, and use of periodicals as the significant correlates of time spent viewing in strings. The variables which remain in this and the prior solution are identical, but there is a different ordering of the primary correlates—for channel changing it was the number of adults in the house, and for string viewing it is the negative relationship with listening activities.

Viewing in mini-stretches.   Here, the time unit was viewing in segments of 5 to 15 minutes. (See Table 5.) The pattern has been established, and the variables which emerge as predictors of this form of viewing will continue to be redundant, given the identified overlap of viewing in this time length with that of even shorter time periods. Adult frequency in the house and age continue to be the sole demographic correlates of viewing behavior; sex and income have no apparent independent role, and number of children and education minor roles at best (with criterion variables other than this one). Mini-stretch viewing is also

Table 4.   Multiple Regressions Predicting Time Viewing in Strings

	R	Beta
Demography		
Age of Respondent	.214	− .219
# of Adults in Household	.295	.203
Environment		
Gadgets in Home	.298	.251
Audio Leisure Activities	.369	− .247
Sports Leisure Activities	.426	.215
(Watching Alone)		(.161)*
Process		
Adventurous Viewing	.359	.359
Overall		
Audio Leisure Activities	.273	− .320
# of Adults	.371	.233
Sports Leisure Activities	.421	.248
Use of Periodicals	.477	− .229

*p = .064

predicted from the environmental set of communication gadgets and sports and audio leisure-time activities. The primary process variable of import is adventurous viewing, although number of favorite channels and the degree of disagreement over how choices are made provide some additional support.

Table 5.   Multiple Regressions Predicting Viewing in Mini-Stretches

	R	Beta
Demography		
# of Adults	.255	.260
Age of Respondent	.331	− .212
Environment		
Gadgets	.316	.276
Audio Leisure Activities	.379	− .231
Sports Leisure Activities	.415	.172
Process		
Adventurous Viewing	.383	.383
(# of Favorite Channels)		(− .158)*
(Upset over Choice Process)		(.157)*
Overall		
Adventurous Viewing	.314	.241
# of Adults	.402	.270
Audio Leisure Activities	.468	− .241
(Gadgets in home)		(.172)*
(Periodical Use)		(− .169)**
(Sports Leisure Activities)		(.165)**

*p < .08, **p < .09.

In the overall solution, adventurous viewing provides the primary relationship with viewing in mini-stretches; number of adults and audio leisure time are also significant. Three other variables—gadgets in the home, use of periodicals, and sports time—approach significance. The first three variables, however, account for 22% of the variance in this viewing behavior.

Viewing in stretches.   Here, the viewing time of interest is lengthened to periods from 15 to 25 minutes in length. The best demographic predictors (see Table 6) are the number of adults and the number of children in the home. Gadgets alone is significant from the set of environmental considerations, but estimated number of TV hours approaches significance as a predictor. Adventurous viewing again is the only significant process variable. Three of these variables remain sturdy in the overall regression analysis, one from each of the variable sets: adventurous viewing, home gadgets, and number of adults. The earlier importance of number of children is better accounted for from among the variables which remain. For only three variables, the multiple R = .438 is notable.

Viewing time in maxi-stretches.   The ability to use these variable subsets to predict longer viewing portions, at least 26 minutes each, is considerably weaker, as evidenced in Table 7. Longer periods of viewing are likely done by people with less education, and in homes with more adults. Self-estimates of total viewing are correlated with electronic measures of household viewing time in longer segments, but no process variables are predictive of this. And only the number of adults in the household is a significant correlate when these variable sets are merged into one predictive set. Note also that, as we have moved to longer time periods, the overall regression solution is weaker. Here, the final multiple R = .286, which identifies 8% of the variance.

Channel changes in maxi-stretch viewing periods.   Here, again, the subsets provide weak associations with this criterion behavior. Table 8 indicates that education, self-estimated television viewing hours, time spent with sporting activi-

Table 6.   Multiple Regressions Predicting Stretch Viewing

	R	Beta
Demography		
# of Adults	.229	.252
# of Children < 18	.308	.207
Environment		
Gadgets in Home	.379	.379
(Estimated TV Hours)		(.142)*
Process		
Adventurous Viewing	.344	.344
Overall		
Adventurous Viewing	.305	.268
Gadgets in Home	.386	.226
# of Adults	.438	.211

*p = .10

Table 7. Multiple Regressions Predicting Viewing Time in Maxi-Stretches

	R	Beta
Demography		
Education of Respondent	.210	−.210
(# of Adults)		(.177)*
Environment		
Estimated TV Hours	.190	.190
Process		
	(no significant correlates)	
Overall		
# of Adults	.286	.286
(Education of Respondent)		(−.173)*

*p < .10

ties, and the extent to which household members are upset during the choice process are separately correlates of channel changing. However, when all subsets are combined, a single variable emerges as indicative of this behavior, and it is the number of adults in the household, with estimated television hours a second possibility.

DISCUSSION

First, the disappointments: In few of the model-testing regression analyses did the viewing style variables make as much of an impact as was hoped for. One—adventurous viewing—remains, from the set utilized, most promising for continued exploration.

Table 8. Multiple Regressions Predicting Total Channel Changes in Maxi-Stretches

	R	Beta
Demography		
Education of Respondent	.189	−.189
(# of Adults)		(.178)*
Environment		
Estimated TV Hours	.197	.243
Sports Leisure Activities	.268	.187
Process		
Upset over Choice Process	.197	.197
Overall		
# of Adults	.280	.280
(Estimated TV Hours)		(.183)*

*p ≤ .08

But, given the primary constraint of the study, that of reflecting household measurement of viewing data rather than individual data, the overall results from this approach are more gratifying than disappointing. First, where adventurous viewing emerged as most significant is where it should have emerged; the variable was comprised of channel flipping during shows to watch more than one show at a time, as well as the pre-viewing behavior of not being definite about what to watch. It became the strongest predictor developed of viewing in intermediate length bursts, in mini-stretches (5–15 minutes) and stretches (16–25 minutes). This would be a stronger statement if it had also been predictive of string viewing; it was, when the set of process variables was taken by itself; but it dropped off in the final set of predictors, largely because of the strong influence of age. In fact, its simple correlation with string viewing (.359) was its highest relationship with any other variable in the study.

Process variables that did not serve as predictors of household viewing (e.g., number of channels checked, number of favorite channels, nonselective viewing) have need to be replicated in a study where individual viewing data are available. Nonselective viewing, e.g., stopping at the first show that looks good, may reflect such a random mode of making television program choices as to be nonpredictive. But we anticipated that it would also have suggested a heavier viewer.

Let us review what has been learned. For a series of household viewing behavior measures of time, channel changing and differential channel selection, a small subset of variables (never exceeding four) accounted for multiple correlations ranging from .44 to .50 for all criterion measures, save those of extended (maxi-stretch) viewing periods. So, measures designed to deal with the new forms of watching television—where remote control and a large array of channels are available—and with activities competitive to television watching accounted for a significant portion of the behavior under scrutiny. The variables which repeatedly assessed these different (sometimes redundant) behaviors were:

1. One demographic characteristic—the number of adults in the house related to heavier viewing.
2. Alternative leisure-time activities—most notably complementary participation in sporting activities and the competitive edge of audio activities. So not all leisure-time activities detract from television viewing time; sports fans are also television fans.
3. Adventurous, active participation by the viewer in trying to follow multiple programs, given no specified roster of programs to watch, is characteristic of heavier viewing of certain types and of the greater channel changing that accompanies those types.

Some criterion behaviors inhibited viewing: the reading of periodicals and the education level of the respondent. For some, the availability of other communication gadgets in the home signaled a more avid and active television fan as well.

When it came to the "traditional" television viewer, he or she who turns to a show and watches it from start to finish, or at least leaves it on for substantially longer periods (at least 26 minutes), the prediction process is much less substantial, the multiple correlations are much lower, and only the number of adults in the household is a useful predictor.

Sex differences, found consistently in other studies in this volume, did not emerge in these data. This is likely due to the prediction of household viewing behaviors, which mixes male and female patterns.

As in other research, it is most difficult to come up with a set of predictors that accounts for this longer form of television time. But again, what this study contributes is an enhanced understanding that total television viewing time includes a substantial chunk of viewing which is not program length (or quarter-hours, as ratings firms would assess it). When this kind of time is added into household viewing estimates, then a stronger prediction is possible (cf. Table 2). Communication gadgetry in the home, and the respondent's education, boost the prediction of total viewing time to a multiple correlation of .468. A more liberal form of regression analysis (looking at all variables simultaneously, rather than the stepwise deletion used here) would have made these results appear even stronger, given the several variables identified in the results as approaching significance in the final estimates presented. But we chose the more conservative route, given the exploratory nature of the study and the imprecise fit between what electronic data was available to us and what would ideally have been available.

Of course, much is left unexplained when the multiple correlations account for about 25% of the variance in the behavior under scrutiny, even when it is a best guess that more would be explained with individual viewing data. Personality assessments are possible, as would be other demographic and environmental characteristics. But primarily, further efforts from these researchers would focus in two directions. First, as one might guess, would be greater elaboration of the viewing-process variables, as they have been elaborated in some of the studies which were fielded after this one. Specifically, there is need to elaborate the *channel changing* that is perceived to be done by the users, e.g., when it occurs and the motives governing it. There is need to elaborate on the users' approach or *orienting behaviors*, how they go about choosing the initial show, beyond the single questions used here. There is need to extend the choice process situation beyond that of procedures used in solo viewing situations, to accommodate what is done when viewing is not done alone, i.e., is *group viewing* and perhaps group decision making, although the latter in itself may require a reconceptualization of much of the process and we will reserve that for the concluding chapter in this volume.

The second path necessary is to re-examine the viewing behavior measures in a manner in which the rating services would be little interested. The present concern ought be with understanding individual or family or household viewing behavior, but, rather than in some absolute or total sense, in the context of time available to those individuals for viewing television. What is proposed is the need to establish, for a given individual, what leisure time is available for allocation. Is it 4 hours or 6 or 8, or what? That, then, would constitute the parameter within which television time, channel hopping, and program flipping should be examined. Is it not more meaningful to attempt to predict what portion of *available leisure time* is given to these television behaviors, rather than be concerned with the total viewing done? Predictions of media-time use have been notoriously weak, yielding low-level multiple correlations of the order obtained here. The suggestion being made is that the target predictor is the wrong one, that focusing instead on attempting to predict time allocation as a function of available time (and corresponding behaviors from the realm of the choice process of central interest here) could be a productive venture.

## CHAPTER 11

Parental Influences on Viewing Style

Carrie Heeter, Bradley S. Greenberg,
Thomas F. Baldwin, Ronald Paugh,
Roger Srigley, and David Atkin

INTRODUCTION

Beginning in the late 1970s, curriculae were developed to teach children critical viewing skills such as evaluation and management of one's own TV-viewing habits and questioning the reality of TV programs (Lloyd-Kolkin, Wheeler, & Strand, 1980; Singer, Zuckerman, & Singer, 1980; Dorr, Graves & Phelps, 1980). Research on cable television brought awareness of viewer selection processes and introduced the concept of *active viewing*. With existing media and with new media, the trend is toward recognition of greater viewer control.

The potential benefits of active viewing make it an area worth pursuing. Active choosing which encompasses evaluation of many alternatives at the time of program selection can maximize a viewer's likelihood of locating the programming best suited to him or her at any given moment. With cable, active searching of alternatives is a prerequisite to taking full advantage of the content offerings available. Once a show has been selected, active viewers may continue to critically evaluate the content, rather than passively accepting all. If it ceases to fill their needs, they may turn elsewhere, in mid-show. Or when commercials come on, the active viewer may choose to turn elsewhere to avoid watching commercials. More than one show can be watched at one time, providing more stimulating and varied content.

Measures to describe an individual's viewing style along these lines have been developed and applied to adult viewers. But issues of how children's viewing styles differ (or do not differ) from adults', and how children's viewing styles develop, have not been addressed. This chapter compares child and adult viewing styles and examines relationships between a parent and child's viewing style.

There is some evidence to suggest that children's viewing styles include extensive sampling. Lyle and Hoffman (1972) describe viewing behavior of first, sixth, and tenth graders, reporting a tendency not to preplan viewing. At all ages, they found that "students were most likely to turn on the set and flip channels to see if there was a program that interested them" (p. 177). Only about one third checked a guide.

Banks (1980) examined Nielsen data on children's viewership, using it to argue that children are less loyal viewers than adults, and are more likely to "rotate among different programs within a given time period" (p. 52). He cites seasonal and week-to-week variations in program ratings as evidence, noting that ratings among younger children (2 to 5 years old) showed even less stability than 6- to 11-year-olds. Banks considers the patterns of varied viewing by children to be "an expression of deliberate search on their parts even at quite an early age" (p. 52).

Chapter 7 presents an analysis of Saturday morning viewing which supports the proposition that children are more frequent channel changers than adults. Households with children evidenced more sampling and extended sampling than other households across three Saturday mornings.

Chaffee, McLeod, and Atkin (1971) examined parental influences on adolescent media use. They measured TV-viewing time, time with specific program types, and time with specific TV shows, correlating parent and child behaviors. They identified three types of influences which could occur: *positive modeling*, where a child engages in the same behavior as the parent; *negative modeling*, where a parent and child both do not engage in some behavior; and *reverse modeling* (particularly appropriate in the realm of television), where a parent actually models the child's behavior. Chaffee et al. (1971) point out that there is an "opportunity factor," which may cause similar parent–child behaviors: availability of content or medium may result in common behavior, rather than being the result of modeling. For example, subscription to cable or availability of a VCR in a household encourages certain viewing styles. Although Chaffee et al. found weak positive correlations between parent and child behaviors, they conclude that "the similarities are not as strong and are at least as amenable to alternative interpretations as they are to the inference that parents directly influence adolescent media use" (p. 324). Their study examined content use and viewing time. The present study will examine viewing style which may be more conducive to modeling.

Marketing research has also examined the concept of consumer socialization, drawing upon Bandura's social learning theory (Ward, 1974). Here, the focus is on children's learning of decision-making processes, which researchers suggest may occur largely by observation rather than specific instruction. The situation is similar with viewing style. Certain of the viewing-style behaviors are observable (e.g., channel changing and guide use), while others are less so (e.g., deciding what to watch before turning the set on, planning an evening's viewing). The directly observable behaviors seem the most likely candidates for modeling.

Viewing-style behaviors will be drawn from our model of program choice. These include orientating search, guide use, planned viewing, re-evaluation, and channel repertoire.

The following research questions are addressed:

Q1:   What is the nature of children's channel repertoires? What is the relationship between the set of channels a child watches regularly and the set of channels his or her parent regularly views?

Q2:   In general, how do children's and adults' viewing styles compare?

Q3:   Within a family unit, how are parent and child viewing styles related?

Q4:   Can a typology of family viewing styles be developed?

METHODS

In-school surveys were administered to 421 fifth and sixth graders at three elementary schools in a medium-sized midwestern town. Students were also given a questionnaire with ID numbers corresponding to their own survey, and a cover letter to take home to their parents to fill out. Students who returned a parent questionnaire in 1 day were given three MSU pencils; those who did so in 2 days were given two pencils, and in 3 days, one pencil. The return rate for parent questionnaires was 87%, with a total sample size of paired parent and child questionnaires of 359.

The child sample was 53% male. Sixty-seven percent of the children reported subscribing to cable television; 32% of the households owned a VCR. Among the parents, 26% of those responding were male. Average annual family income ranged between $35,000 and $45,000.

The cable system in the sample community carried 43 channels, including six pay channels. Three network affiliates, a PBS station, three independents, and a Canadian station were available over the air.

Questionnaire

The same questions were asked of children and parents. For each channel available to respondents (a different set for cable and noncable households), the children were asked to circle those which they watched at least once a week. Parents were asked to indicate the number of days per week they watched each channel. For a directly comparable measure of channel repertoire (the number of different channels watched regularly), parental responses for each channel were collapsed into two parallel categories: not watch, and watch once or more per week. The total number of channels watched was summed.

Two items addressed orienting search: how often respondents checked many channels, then went back to the best one; and how often they stopped at the first show that looked good. For re-evaluation, respondents were asked how often they changed channels and how often they checked channels to see what else was on when the show they were watching ended, at commercials during a show, and during a show not at commercials. Frequency of watching more than one show at a time was also asked.

Two guide-use items differentiated between frequency of guide use before viewing and frequency of use while viewing. Six items assessed respondents' tendencies to plan viewing. How often respondents watched the same shows as the weekday before and as the weeknight before was assessed, as well as knowing what show to watch before turning on the set, knowing what show to turn to next when the show being watched ends, checking channels when one show ends, and watching whole shows from start to finish.

Parents were asked how often they and how often their fifth- or sixth-grade child got to decide what to watch, when watching together. Children were asked the same things. Each was asked how frequently the other changed channels, and the frequency with which someone else changed channels when they themselves wished they wouldn't.

Analysis

First, adult and child viewership of individual channels among cable and noncable households are compared using chi square to test the significance of the relationship. For each channel, a 2×2 contingency table is constructed with axes of parent viewership (yes/no) and child viewership (yes/no). Means on media-use and viewing-style variables are compared using paired t-tests, and simple correlations calculated. Finally, the adult and child measures are subjected to canonical correlation analysis to link parent and child behaviors.

RESULTS

Channel Repertoire

Table 1 shows the percentage of children and parents who identified each channel as belonging to their repertoire of regularly-viewed channels. Cable and noncable households are reported separately because they have different sets of channels available. On the average, children in cable households regularly watched an average of 11 channels, while those without cable watched an average of five. Among parents, an average of nine channels were watched among cable households, and four among noncable. Children reported watching more different channels than parents in both environments.

Of the 18 channels watched regularly by one-fourth or more of the sample cable children, 11 were cable-only channels. These included five pay movie channels: Nickelodeon, USA, ESPN, MTV, and superstations WTBS and WGN. Children who had cable were somewhat less likely to include the broadcast net-

Table 1. Channel Repertoire

Channel	Cable Households		Noncable Households	
	Child %	Adult %	Child %	Adult %
HBO	82%	81%**		
WKBD (Ind.)	78%	58%	88%	49%
WDIV (NBC)	76%	75%	92%	76%
WXYZ (ABC)	75%	85%	93%	79%
WJBK (CBS)	68%	72%	89%	73%
WXON (Ind.)	59%	34%	68%	32%
The Movie Channel	55%	36%**		
Nickelodeon	55%	36%**		
USA Network	51%	29%*		
Cinemax	46%	24%**		
ESPN	45%	45%**		
WGN (Chicago)	42%	37%**		
WTBS (Atlanta)	41%	29%**		
MTV	41%	16%**		
CBET (Canada)	28%	23%	30%	17%
Showtime	26%	11%**		
The Disney Channel	26%	11%**		
WTVS (Public TV)	25%	38%	27%	29%
The Weather Channel	23%	23%		
Weather-radar	18%	24%		
CNN	17%	23%		
PASS (Sports)	15%	5%		
CNN Headline	15%	20%*		
WGPR (Ind.)	10%	8%	12%	6%
Omni-8 (local)	10%	2%		
SPN	8%	2%		
CBN (Christian)	8%	8%		
Local Schools	8%	14%		
Local News	7%	12%		
Home Theater Network	7%	2%		
Program Guide	6%	5%		
Metro (local)	5%	0%		
Nashville	5%	3%		
Lifetime	5%	5%*		
Schoolcraft College	5%	3%		
MSN	5%	0%		
Library	4%	1%		
Local Government	3%	5%		
CAP-15 (local)	3%	3%		
Financial News Network	3%	2%		
AP News	3%	0%		
C-Span	3%	3%*		

**Indicates significant χ^2 association at p < .005; *indicates p < .05

works and independent stations in their channel repertoire, although the relative popularity of broadcast channels was similar between cable and noncable children.

Chi square tests of association between parent and child channel use were run for each channel among cable and noncable households. In noncable households, none of the tests of parent–child association were significant. In cable households, for 15 channels, chi square tests of the overlap of parent and child channel repertoires were significant. All 15 shared channels were cable-only channels. Not surprisingly, five of the 15 significantly shared channels were pay movie channels. (Because the movie channels must be subscribed to to be viewed, one would expect a tendency for subscribing household members to use those channels in common.) For the 10 other significantly shared cable-only channels, the issue is how those channels came to be shared. The direction of influence is likely two-way. Some of the shared channels would probably be initiated by adult preference and viewing, which the child then adopts (e.g., The Weather Channel, CNN Headline, Lifetime, and C-Span). This fits Chaffee et al.'s (1970, 1971) definition of modeling. Others may more likely be initiated by the child (Nickelodeon, MTV, USA Network) fitting the pattern of reverse modeling. ESPN, WGN, and WTBS could conceivably reflect either child or parental preferences and influence.

Data analysis for media-use and viewing-style variables addresses two issues: how adult and child self-reported behaviors differ, and what the relationship exists between children's and their parent's behaviors. Thus, the first two columns of Tables 2 and 3 report aggregate child and adult means for each behavior. Column three presents Pearson's correlations of child and parent responses.

Media Use

Consistent with expectations for this age group, fifth and sixth graders spent more time with television, went to more movies, spent more time with records/

Table 2. Media Use Means and Simple Correlations

	Child Mean	Adult Mean	r
Daily TV viewing time	4.25 hours	*3.25 hours	.14*
Daily newspaper viewing time	0–10 minutes	*2.6 minutes	.10*
Days/week read newspaper	2.3 days	*5.8 days	.15*
Magazines read last month	1.5	*2.0	.11*
Nonschool books read/month	2.4	*1.7	.03
Movies seen last month	1.7	*0.9	.26*
Daily radio hours	2.4	*2.8	.01
Daily record hours	1.8	*0.8	.02
VCR use[1]	(1.3)	*(0.5)	.31*

[1]children were asked days per week of VCR use, and parents hours per week
*indicates significance <.05 for paired t-test mean comparisons and correlations

Table 3.   TV Viewing Style Means and Simple Correlations

	Child Mean	Adult Mean	r
Channel repertoire			
(cable households)	11.1	*9.0	.38*
Channel repertoire			
(noncable households)	4.9	*3.6	
ORIENTING SEARCH			
Stop at first good channel[1]	0.8	*1.4	−.02
Check many, go back to best	0.8	*1.4	.13*
RE-EVALUATION			
Check chans when show ends	1.1	*1.6	.16*
Check chans at commercials	1.5	*2.0	.13*
Check during show	1.9	*2.2	.16*
Watch >1 show at a time	1.8	*2.4	.14*
GUIDE USE			
Check guide before watching	1.6	*0.8	.15*
Check guide while watching	1.7	*1.1	.02
PLANNING			
Watch same weekday shows	0.8	*1.6	.01
Watch same weeknight shows	0.8	*1.1	.16*
Plan shows before viewing	0.8	0.8	.02
Know next show when one ends	1.0	*1.1	.02
Watch shows start to finish	0.8	0.7	.14*
CONTROL			
You choose with parent/child	1.5	*1.0	−.15*
Parent/child changes channels	1.6	1.4	−.02
Others change when not wanted	1.2	*1.7	.14*
Watch with siblings/others	1.5	*0.7	.07

[1]for the remaining variables, response categories ranged from 0 (very often) to 3 (not very often)
*indicates significance <.05 for paired t-test mean comparisons and correlations

tapes, and read the newspaper less often than their parents (see Table 2). The children read more books and fewer magazines. Radio listening occupied more than 2 hours for both groups. Consistent with prior research, significant but weak positive correlations were found between parent and child TV viewing and newspaper readership as well as magazine reading. Children whose parents saw more movies in the theater were themselves likely to see more movies. VCR usage was positively related (r = .31), no doubt due to the availability of the equipment in the home.

Viewing Style Comparisons

Viewing-style comparisons are reported in Table 3. Children reported larger channel repertoires than their parents, in cable and noncable homes. This variable showed the largest parent–child correlation of any of the media behaviors examined (r = .38).

On the average, children and adults reported engaging in terminating orienting searches (stopping at the first show that looks good) with equal frequency as exhaustive searches (checking channels and going back to the best option), although the fifth and sixth graders did more of both behaviors than adults. Terminating-search behavior was uncorrelated between parent and child, while exhaustive searching was moderately correlated ($r = .13$).

For each of the four re-evaluation items (checking channels during a show, at commercials, between shows, and watching more than one show at a time), children reported more frequent re-evaluation behavior than the parents. Positive, consistent, significant parent–child correlations emerged for each re-evaluation variable.

Fifth and sixth graders used a TV guide less often than parents. Checking a guide before watching television was significantly correlated between parent and child, although guide use while watching was not.

Children were more likely to watch the same weekday and weeknight shows as parents. However, they were also more likely to check channels to see what else was on when a show ended. Planning what to watch before turning on the set, and watching a show from start to finish, were reported equivalently by adults and children. Only two planning variables were significantly correlated within a household: watching the same weeknight shows, and watching a show from start to finish.

Parents were more likely to indicate that *they* selected shows when watching with children. Parental reports of the frequency with which their child changed channels, and child reports of parent channel changing, were equivalent. The children more often indicated that someone else changed channels when they wished they wouldn't. The children also watched television more often alone. Parent–child estimates of which party selects programs to watch were significantly correlated. In addition, children who claimed someone else changes channels when they wished they wouldn't were more likely to have parents who claimed the same thing.

Canonical Correlation

The 18 parent and child media-use and viewing-style items which showed significant correlations were entered into a canonical correlation analysis. With listwise deletion of missing values, the sample size included in the canonical analysis was 315 parent–child pairs. Three significant roots emerged. Table 4 reports their respective canonical correlations, χ^2 significance levels, and parent–child variables loading on the roots. The three significant canonical roots combined could account for 59% of the variance between parent and child media behaviors.

Canonical root 1 explained 33% of the variance, and was labeled *Large Repertoire*. On the basis of the variable loadings on the canonical root, a household type can be characterized.

Table 4. Canonical Correlations Results (n = 315)

Root 1: Large Repertoire
Canonical correlations = .573 $\quad\quad\quad\quad\quad\quad\quad\quad$ χ^2 p = .000

Child Variables	Adult Variables
.56 Channel Repertoire	.50 Channel Repertoire
.38 Same Weeknight Shows	.37 Same Weeknight Shows
.33 TV Time	−.35 Others Change Channels
.28 Watch Shows Start to Finish	−.32 Check Channels When Show Ends
	.30 Watch Shows Start to Finish

Root 2: VCR Households
Canonical correlations = .501 $\quad\quad\quad\quad\quad\quad\quad\quad$ χ^2 p = .000

Child Variables	Adult Variables
.77 VCR Viewing	.72 VCR Viewing
−.44 Check Channels During Show	−.39 You Choose What to Watch
.42 Check Channels at Commercials	
.33 Watch More Than 1 Show	

Root 3: Channel Checkers
Canonical correlation = .425 $\quad\quad\quad\quad\quad\quad\quad\quad$ χ^2 p = .021

Child Variables	Adult Variables
−.47 Others Change When You Wish they Wouldn't	−.73 TV Viewing Time
.39 Magazine Reading	−.62 Watch More than 1 Show
.30 Check Channels During Show	.51 Check Many Channels, Then Go Back
.33 Check Many Channels, Then Go Back	−.45 Check at Commercials
.30 Check Channels When Show Ends	.42 Check Channels During Show

In large repertoire households, both parent and child report that they regularly watch many channels. Also directly parallel between parent and child are the tendency to watch the same weeknight shows each week and to watch shows from start to finish. Children in these households watch more television than other children. Parents indicate that someone infrequently changes channels when they wished they wouldn't and that they themselves often watch shows from start to finish. Although there is a large channel repertoire, there is little indication of much channel changing.

Root 2, labeled *VCR Households*, accounted for another 17% of the variance. In VCR households, parents and children report frequent use of VCRs. Children in the VCR household tend not to check channels during a show, but they do check channels at commercials and watch more than one show at a time. Other than watching VCRs, parents in these households are distinguished by not themselves choosing what the group will watch.

Root 3, labeled *Channel Checkers*, accounted for an additional 9% of parent–child viewing-style variance. Two variables characterize both parent and child in these households: checking many channels, then going back to the best one and checking channels during a show. The parents do not watch much television. They tend not to watch more than one show at a time or to check channels at commercials. The children do also check channels when a show ends. Children do not report that someone else changed channels when they wished they wouldn't. They tend to read more magazines.

CONCLUSIONS

Parent and child viewing styles are definitely related. Simple correlations tend to be small but significant. However, canonical correlation analysis was able to explain 59% of the variance in parent–child viewing styles. Children's channel repertoires were consistently higher than adult repertoires, by one to two channels. The fifth and sixth graders do watch more television than their parents, but these data indicate they also derive that content from more different channels. It is unlikely that there are more cable channels designed to appeal to children than to adults. What is more likely, perhaps, is that children look harder for programming. Child repertoires include most of the cable channels that regularly carry children's programs. Whether this parent–child difference will continue when the first generation of "cable kids" become parents remains to be seen.

The difference in shared parent-child channels between broadcast and cable-only channels is striking. Not one broadcast channel, compared to 15 cable-only channels, was significantly related within a parent–child pair. This is likely a result of the larger set of cable channels, greater specialization of content, and greater availability of children's programming.

The fifth and sixth graders did report more re-evaluation, channel switching, and channel checking during orienting searches. They used guides less frequently. Planned viewing was not different. These measures are self-report data. Actual behavioral measures of parent and child viewing style could help differentiate measurement artifacts from behavioral differences.

Movie going, VCR use, and channel repertoire were the three media behaviors most strongly correlated between parent and child. Opportunity factors are certainly operating here—in terms of chances to attend movies, availability of a VCR, and availability of cable. Modeling and reverse modeling are both possible, and further study should attempt to determine which operates to what extent for these variables.

The intervening influence of parental mediation should be assessed. The role and relative impact of cable subscription should also be isolated in further studies. To date, parental mediation variables have not focused on viewing style, but a special set could be developed.

The canonical correlation analysis identifies one typology of parent–child viewing style. Large-repertoire households reflected a positive orientation toward television, watching shows from start to finish, repeatedly viewing the same weeknight shows, and watching many different channels. VCR household viewing was associated with watching shows on videocassette, watching more than one show at a time, and zapping commercials—a more aggressive, active approach to viewing. Channel checkers were also active, checking many channels and going back to the best one, and checking channels during a show. However, they tended not to check channels at commercials or watch more than one show. They watch less television. When they do watch, it is as if they change channels to seek something which they are not finding, rather than changing channels to maximize the amount of TV content they can watch, like the VCR households.

This typology suggests a new characterization of viewing style. Other than this study, our book has described viewing style as almost unidimensional and linear, with some individuals almost never changing channels and others almost constantly changing channels. The canonical correlation analysis integrates more viewing behaviors and, significantly, VCR usage. The result is perhaps a more sensitive characterization of viewing styles.

The existence of the household types identified through canonical correlation should be replicated and refined. Factor analysis among adults only using the same variable set should be run.

CHAPTER 12

Gender Differences in Viewing Styles

Carrie Heeter

Previous chapters report data which suggest that women and men approach television viewing differently. This chapter gathers data across 10 studies to definitively document and describe differences in female and male viewing styles. The studies were conducted between 1982 and 1985. A total of 3,168 adults and 1,899 youth were surveyed. Methods ranged from door to door to telephone to in-school to take-home questionnaires. Respondents were drawn from nine cities in five states. Cable systems carried between 36 and 50 channels.

Table 1 summarizes the 10 studies, presenting a descriptive name (e.g., Cable Adult 1) which will be used to refer to that study throughout this chapter, the year it was conducted, overall sample size and the number of female and male respondents, data collection method, an alphabetic identifier of city (to be described later), the proportion of cable penetration among sample respondents, the number of channels carried on the local cable system, and the book chapter in which the study is described in greater detail (where applicable). A brief description of the study also appears on Table 1.

Four of the studies (Cable Adult 1, Cable Adult 2, Mixed Adult 1, and Mixed Adult 2) were conducted in a medium-sized, midwestern city with cable systems carrying 36 channels. Four of the studies (Mixed Parent 1, Mixed Parent 2, 5th Grade and 10th Grade 1) were conducted in an upscale midwestern community with a 50-channel cable system. Playboy respondents were randomly selected from Playboy Channel subscriber lists in four cities from four different states. Tenth Grade 2 respondents were drawn from schools in three medium-sized midwestern cities with approximately balanced ratios of Blacks and Whites.

Three studies surveyed only respondents who had cable (Cable Adult 1, Cable Adult 2, and Playboy). Cable Adult 1 was a door-to-door survey, in which a screening question about whether or not the household subscribed to cable was the basis for selecting cabled households, using geographic clusters with a fixed-skip interval and randomized start point. Cable Adult 2 consisted of three waves

Table 1.   Description of Studies

Study:	Year:	Sample Size: total: f/m	Method:	City:	% Cable	# Channels:	Book Chapter:
Cable Adult 1	1983	230:109/121	door to door	A	100%	36	2
random cable subscribers, by geographic area							
Cable Adult 2	1985	1428:756/672	phone	A	100%	36	8
random cable subscribers from subscriber list, equal distribution of basic, pay, multipay							
Playboy	1984	386:79/207	phone	B	100%	36+	15
random Playboy Channel subscribers (from subscriber lists) from 4 cities in 4 states							
Mixed Adult 1	1982	339:163/167	phone	A	62%	36	16
fixed-skip interval from area phone book plus randomized last digit							
Mixed Adult 2	1983	254:115/139	phone	A	61%	36	—
fixed-skip interval from area phone book plus randomized last digit							
Mixed Parent 1	1984	180:112/58	written	C	67%	50	—
parents of 10th graders who returned take home survey (46% completion)							
Mixed Parent 2	1984	351:261/90	written	C	67%	50	11
parents of 5th graders who returned take home survey (86% completion)							
5th Grade	1984	407:193/214	written	C	67%	50	11
in-school survey of 5th graders							
10th Grade 1	1984	395:180/215	written	C	64%	50	—
in-school survey of 10th graders							
10th Grade 2	1985	1097:529/568	written	D	78%	36+	—
in-school surveys of 10th graders from three cities with racially balanced populations							

of telephone data collection, in conjunction with the design and delivery of a free, localized cable-viewing guide. Respondents were randomly selected from subscriber lists to receive the guide and be surveyed. The first wave of data collection occurred prior to introduction of the guide. Wave 2 data were collected after the monthly guide had been distributed for 3 months. Wave 3 data were collected after the guide had been distributed for 6 months. Equal numbers of basic, pay, and multipay subscribers were selected. Playboy respondents all subscribed to the Playboy Channel in addition to basic cable and, usually, other pay channels.

Mixed Adult 1 and Mixed Adult 2 used fixed-skip intervals from area phone books plus randomized final digits to survey cable and noncable respondents in proportion to their actual distribution in the population.

Mixed Parent 1 and 2, 5th Grade and 10th Grade 1 involved in-school administration of surveys to youth, plus a take-home questionnaire for the students' parents. The parental return rate for 5th grade parents (Mixed Parent 2) was 86%, compared to 46% for 10th grade parents (Mixed Parent 1). Only one parent responded to each survey—thus, either the father or the mother answered, not both. A larger proportion of mothers responded at each grade level. Tenth Grade 2 also involved in-school surveys, with no parallel parental take home instrument.

Respondents in all the studies are predominantly cable subscribers, from a low of 62% to a high of 100% in cable only samples.

Many viewing-style questions asked were common across surveys. However, given the wide range of populations studied, it does not make sense to combine the data into one massive database. The result would be a nonrandom mishmash, unrepresentative of any population. Instead, female–male comparisons will be reported for each of the studies separately, permitting inferences about possible differences across studies on the basis of the populations. For example, three of the studies are of youth, while seven are of adults. If results for adults are significantly different for females and males, while results for youth are not, one may posit a confounding impact of age on viewing-style gender differences.

Many of the tables present female–male mean comparisons for viewing-style questions which were asked across some combination of the 10 studies. The parent and child surveys for 5th and 10th graders used four-point instead of five-point scales because fewer response categories were considered more appropriate for that age group. To make these results more comparable with the five-point scale results, the means were multiplied by 1.25, yielding the equivalent of a five-point scale. Three of the five-point scale studies were based on the following response categories: almost always, 3/4 of the time, 1/2 the time, 1/4 of the time and almost never. The remaining two five-point scale studies were based on these options: very often, quite often, often, not very often, and not at all. In each table, 4 indicates almost always (or very often) and 0 indicates almost never (or not at all). Occasionally, other response categories are used. Text and tables will so indicate. T-tests were conducted within each study to assess probability levels for observed female–male differences.

RESULTS

The results are organized into four sections: orienting search and planning, guide use, reevaluation, and viewing style by program type.

Orienting Search and Planning

Table 2 reports *channel changing when respondents first turn the TV set on*. For four of five adult studies (all but Playboy), males change channels more often. Youth differences are not significant.

Table 2 also supports the claim that adult females *know what they are going to watch before they turn the set on* more often than do adult males. For three of six studies, the results are significant. Two others approach significance ($p = .062$ and $p = .100$). Even for Playboy subscribers, the direction is consistent. This difference is not evident for youth samples.

Table 3 taps *repeated viewing of the same shows for daily and weekly programs*. There is a strong tendency among adults and youth for females to watch

Table 2.   Planning

Study	F	M	p	F	M	p
	How often do you check channels when you turn the set on?			How often do you know what you are going to watch before you turn the set on?		
Cable Adult 1	1.51	1.88	.001	2.91	2.56	.062
Cable Adult 2	—	—	—	2.73	2.43	.000
Playboy	1.14	1.30	.270	2.18	1.98	.276
Mixed Adult 1	1.19	1.61	.000	2.88	2.57	.042
Mixed Parent 1	1.53	2.06	.000	2.80	2.50	.038
Mixed Parent 2	1.60	2.01	.001	2.78	2.59	.100
5th Grade	2.65	2.75	.281	2.78	2.79	.842
10th Grade 1	2.66	2.73	.414	2.48	2.34	.178
10th Grade 2	2.29	2.41	.074	2.80	2.74	.352

the same shows they watched the day before more often than males. Youth evidence higher levels of repeat daily viewing than adults. For repeated viewing of weekly shows, the trend is in the same direction, but only one of two adult-study and one of three youth-study mean differences are significant. Parents repeat view weekly shows moreso than daily shows, while youth report approximately the same levels for each.

Two adult studies asked four items measuring viewership of regular shows. All gender mean comparisons are significant at $p < .001$. (See Table 4.) Females more often report that there are *daily shows they watch almost every day*. These same differences were found for *weekly shows they watch almost every week*.

Orienting search process questions are reported in Table 5. Two approaches to channel changing were identified through personal interviews: *changing channels and stopping at the first show that looks good; and checking many channels, then going back to the one that looks best*. Where significant differences emerge, they are consistently in the direction of males doing more of both behaviors than females.

Adult females and males *check channels in order (2 then 3 then 4 . . .)* about

Table 3.   Repeat Viewing

Study	F	M	p	F	M	p
	How often do you watch the same shows you watched the day before?			How often do you watch the same shows you watched the week before?		
Cable Adult 2	2.54	2.26	.001	—	—	—
Mixed Parent 1	1.96	1.53	.027	2.66	2.18	.006
Mixed Parent 2	1.78	1.63	.363	2.40	2.24	.193
5th Grade	2.89	2.73	.100	2.79	2.67	.286
10th Grade 1	2.60	1.95	.000	2.46	2.10	.002
10th Grade 2	2.79	2.42	.000	2.79	2.73	.460

Table 4.   Regular Shows and Switching

(n)	F (109)	M (121)	F (115)	M (139)
Study	Cable Adult 1		Mixed Adult 2	
Are there any weekly shows that you watch almost every week? (% yes)	77%	59%	69%	48%
Are there any daily shows that you watch almost every day? (% yes)	73%	52%	58%	36%
When you are watching *your regular shows*, about how many times per half-hour would you estimate that you change to a different channel?	0.8	2.3	1.0	2.3
When you are watching something *other than a regular show*, about how many times per half-hour would you estimate that you change to a different channel?	1.6	4.2	2.3	4.2

all differences are significant by t-test at $p < .001$.

as often. (See Table 6.) Males with cable are more likely to *check channels not in numerical order*. That difference is not evident among Playboy subscribers. Only three studies asked these questions. Tenth Grade 2 was the only youth study where these questions were asked. For both items, significant tenth grade gender differences were found. Tenth-grade females are more likely to check channels in numerical order, while tenth-grade males are more likely to check channels not in numerical order.

Table 7 reports *search repertoire and channel repertoire* for females and males across three studies. In each study, males claim to check more different channels than females. Cable Adult 1 means are much higher (by 9–12 channels)

Table 5.   Search Method

Study	F	M	p	F	M	p
	How often do you change channels and stop at the first show that looks good?			How often do you check many channels, then go back to one?		
Cable Adult 1	1.28	1.74	.021	—	—	—
Cable Adult 2	1.89	2.00	.207	1.72	2.06	.000
Playboy	1.78	1.87	.639	1.74	1.75	.987
Mixed Adult 1	1.19	1.43	.127	—	—	—
Mixed Parent 1	1.96	2.06	.513	1.88	2.24	.024
Mixed Parent 2	1.93	2.13	.098	1.99	2.18	.118
5th Grade	2.69	2.76	.453	2.83	2.75	.473
10th Grade 1	2.66	2.73	.494	2.53	2.41	.734
10th Grade 2	2.42	2.36	.356	2.49	2.62	.054

Table 6.   Channel Checking Order

Study	F	M	p	F	M	p
	How often do you check channels in order (2, then 3, then 4)?			How often do you check channels not in numerical order?		
Cable Adult 2	1.54	1.56	.898	1.20	1.52	.000
Playboy	1.19	1.43	.430	1.25	1.34	.610
10th Grade 2	2.25	2.09	.034	1.54	1.70	.031

than Cable Adult 2. This is likely due to different measurement techniques. For Cable Adult 2, respondents were simply asked to estimate, by phone, how many channels they checked. For Cable Adult 1, respondents were presented, in person, with a graphic representation of the system's channel selectors, consisting of three rows of 12 buttons labeled with channel numbers in the positions that they appeared on the cable system. After identifying known channels by name, respondents were asked to describe, by pointing, which channels (both known and unknown) they typically checked when they checked channels.

Adult females and males did not report different-sized repertoires of channels watched regularly. Tenth-grade males claimed to watch more different channels than tenth-grade females.

Guide Use

Two guide use questions were asked across numerous studies: frequency of guide use *before* turning on the TV set and frequency of use while watching TV. (See Table 8.) In two of four adult studies, females were found to check a guide significantly more often when they first turn the set on. The two parent studies' results are not significant. Perhaps parents, particularly fathers who fill out children's take-home questionnaires, have different guide behaviors than adults in general. At the 5th-grade level, females check guides more than males before turning the set on, although this difference is not apparent in 10th grade.

Table 8 also presents mean comparisons for *frequency of guide checking while watching TV*. The adult female–male differences which were found for checking

Table 7.   Search and Channel Repertoire

Study	F	M	p	F	M	p
	When you check channels, how many channels do you check?			How many channels do you watch regularly (at least once a week)?		
Cable Adult 1	15.98	20.02	.000	6.41	7.31	.116
Cable Adult 2	7.12	8.40	.000	10.72	11.09	.210
10th Grade 2	10.32	12.75	.001	6.78	8.59	.000

Table 8.   Guide Use

Study	F	M	p	F	M	p
	How often do you check a guide before turning the set on?			How often do you check a guide while you are watching TV?		
Cable Adult 1	—	—	—	1.37	1.29	.676
Cable Adult 2	2.33	1.77	.012	—	—	—
Playboy	—	—	—	3.74	3.94	.161
Mixed Adult 1	2.14	1.81	.050	1.52	1.57	.750
Mixed Parent 1	2.36	2.24	.493	2.14	2.29	.369
Mixed Parent 2	2.74	2.71	.785	2.31	2.43	.312
5th Grade	2.65	2.75	.281	1.76	1.50	.028
10th Grade 1	2.66	2.73	.414	2.25	2.06	.084
10th Grade 2	2.29	2.41	.074	2.01	2.14	.080

a guide before turning the set on disappear. Playboy subscribers (who tend to subscribe to several other pay channels) check guides while watching dramatically more often than other samples. Parents check guides more often than most other adults. As with guide checking before turning the set on, female 5th graders check guides significantly more often while watching than male 5th graders. Tenth Grade 1 results approach significance ($p = .084$) in the same direction, and 10th Grade 2 approaches significance in the reverse direction. In general, youth check guides less frequently than their parents.

Cable Adult 2 included many detailed questions about guide use. Gender comparisons have been analyzed for guide-use advance planning of what to watch, frequency of guide use at various times while viewing, and types of information sought when guides are checked. Females are significantly more likely than males to use a guide *to plan in advance what to watch for the week*. (See Table 9.) Gender differences were not found for *planning in advance what to watch for one day or evening*, although guides were more frequently used for this purpose by both sexes. Gender differences were also not found for guide use at three different periods during viewing. The most frequent time viewers of either sex check a guide is *when they first sit down to watch TV*. Next most frequent is *between shows, when the show they are watching ends*. The least frequent time to check a guide is *in the middle of a show*. Thus, unlike channel changing, female–male differences in guide use are quite limited.

Table 10 reports the percentage of female and male respondents who are trying to find nine different kinds of information when they look in a guide. More females than males seek seven of the nine kinds of information. *Show start time and show name* are sought equivalently by females and males. Females are significantly more likely than males to read a guide to find out *what a show is about, how long a show is, and what stars are on a show*. Significance levels for four other kinds of information range from .060 to .164: *what channel a show is on, guests on a show, whether a show is a repeat, and what other shows are on*.

Table 9. Guide Use

	(from Cable Adult 2 Study)		
	F	M	p
How often do you use a guide to plan in advance what shows you will watch *for the week*?	1.50	1.36	.052
How often do you use a guide to plan in advance what shows you will watch for one day or evening?	1.74	1.68	.459
When you *first sit down to watch TV*, how often do you check a guide?	1.65	1.60	.414
How often do you check a guide *between TV shows*, when the show you are watching ends?	1.30	1.24	.381
How often do you check a guide *in the middle of show* you are watching, to see what else is on?	0.88	0.91	.601

These differences suggest that females invoke more criteria in selecting shows than males do.

This tendency is supported by related results from another study. Mixed Adult 1 asked respondents how often they searched for four different program characteristics when they wanted to watch TV, but didn't have a particular show in mind. (See Table 11.) The criterion most commonly looked for by female respondents was *a series that's familiar*, significantly more often sought by females than males. Females were also more likely than males to look for *an actor/ actress they recognized*, although this was the least frequent search criterion among both sexes. Males most commonly sought *a specific kind of show*, although females were equivalently likely to seek this. Females and males were also equivalently likely to report looking *at some favorite channels* when search-

Table 10. Information Sought from Guides

	(from Cable Adult 2 Study)		
	F	M	p
When you look in a guide for information about a show you might want to watch, what are you usually trying to find out?			
channel show is on	91%	85%	.060
show start time	89%	90%	.583
what other shows are on	85%	80%	.164
what show is about	84%	74%	.008
show name	66%	66%	.922
how long show is	63%	47%	.000
whether show is a repeat	58%	51%	.130
guests on show	45%	37%	.092
stars on show	44%	34%	.022

ing for something to watch. Females considered all criteria except "to look at some favorite channels" more often than males, significantly so for two of three criteria.

Re-evaluation

Tables 12 and 13 report four questions designed to assess respondents' tendencies to re-evaluate a program choice while the show is still on. The first item, *"how often do you change channels between shows,"* has significantly higher male than female means for every adult study. The 1982 study (with the lowest proportion of cable subscribers) has unusually low means, but the gender difference is consistent. On the other hand, neither the 10th- nor 5th-grade gender differences are significant. Further, all of the child means are higher than any of the adult means. Perhaps today's youth are more oriented toward changing channels than adults. Perhaps the gender differences will disappear among adults when these youth grow up, or perhaps the differences will develop as they mature.

The next question addresses zapping of commercials (*how often do you change channels when commercials come on, during a show*). The same basic trends are found. For five of the six adult studies (all except Playboy), males change channels significantly more often. Youth sex differences are again not significant. Youth report changing more often at commercials than adults do.

The least frequent time to *change channels is during a show, not at commercials*. (See Table 13.) Again, males change significantly more than females in every adult study, and, for this item, in two of the three child studies. For 10th Grade 1, the difference is significant but the direction is reversed. These 10th-grade girls report changing in the middle of a show more often than 10th-grade boys. But in a larger sample of 10th graders across three communities (10th Grade 2), the opposite significance is found, consistent with other results. Changing channels during a show, not at commercials, has consistently been the most extreme form of channel changing, reportedly occurring less often than changing at other times. Those who change regularly during a show have been

Table 11.   What People Look for When Changing Channels

	(from Mixed Adult 1 Study)		
	F	M	p
When you want to watch TV and don't have a particular show in mind, how much of the time do you . . .			
look for a series that's familiar to you	2.44	1.87	.000
look for a specific kind of show	2.36	2.23	.415
look at some favorite channels	1.62	1.77	.339
look for an actor or actress you recognize	1.34	1.02	.021

Table 12.   Re-evaluation

Study	F	M	p	F	M	p
	How often do you change channels between shows?			How often do you change channels when commercials come on, during a show?		
Cable Adult 1	1.40	1.88	.013	0.88	1.77	.000
Cable Adult 2	1.90	2.07	.037	1.27	1.69	.000
Playboy	—	—	—	1.14	1.30	.270
Mixed Adult 1	0.68	1.36	.001	0.46	0.87	.000
Mixed Parent 1	1.85	2.15	.038	1.25	1.71	.005
Mixed Parent 2	1.83	2.05	.033	1.15	1.46	.022
5th Grade	2.54	2.66	.262	1.74	1.94	.062
10th Grade 1	2.40	2.36	.734	1.99	2.04	.693
10th Grade 2	2.58	2.60	.684	2.03	2.12	.280

described as the "die-hard" zappers. It is interesting that, on this most extreme measure, significant youth gender differences emerge.

Finally, some support is found (two of four studies) for females to report *watching an entire show without changing channels* more often than males. Where significance is not achieved, the direction is consistent. Youth gender differences are not significant.

Two adult studies asked respondents to estimate how many times they changed channels. (See Table 4.) Respondents were first asked *about how many times per half hour they would estimate they changed channels while watching their regular shows*. The male mean estimate for both studies was 2.3 times per half hour, significantly higher than the female means of 0.8 and 1.0. *Estimates of channel changing per half hour while watching something other than a regular show* were also collected. Male means across the two studies were again iden-

Table 13.   Re-evaluation

Study	F	M	p	F	M	p
	How often do you change channels during a show, not at commercials?			How often do you watch an entire program without changing channels?		
Cable Adult 1	0.32	0.75	.001	—	—	—
Cable Adult 2	0.80	1.06	.000	2.98	2.75	.001
Playboy	0.76	1.01	.032	2.71	2.51	.250
Mixed Adult 1	0.35	0.79	.000	—	—	—
Mixed Parent 1	0.85	1.21	.001	2.91	2.55	.008
Mixed Parent 2	0.91	1.11	.017	2.86	2.75	.246
5th Grade	1.23	1.41	.050	2.84	2.71	.181
10th Grade 1	1.29	1.10	.043	2.54	2.52	.954
10th Grade 2	1.19	1.33	.032	2.81	2.71	.168

tical at 4.2 times per half hour, compared to 1.6 and 2.3 times reported by female respondents. Both sexes change channels more often during shows they do not watch regularly. But in both situations, males change considerably more often.

Mixed Adult 2 asked respondents how frequently they changed channels for particular purposes. (See Table 14.) The most common reason for changing channels, *to turn to a specific channel*, was not different by gender. Males were significantly more likely to change channels *to see what else is on*. There was a tendency approaching significance (p = .066) for males also to change more often *because of boredom*. *To avoid commercials and to watch more than one show* were not significantly different by gender, although the direction was for males to report those motivations more often.

Watching more than one show at a time is a more frequent adult male than female pastime. (See Table 15.) The exception is female Playboy subscribers, who actually watch two shows at the same time more often than male Playboy subscribers. Playboy subscribers watch two shows at the same time more often than other adult groups. Youth watch more than one show at a time more often than their parents. In 5th grade, males do so more than females, but that difference disappears in tenth grade.

Table 16 reports comparisons of perceived control over what respondents watch. Adult females report significantly less *control over what they watch* than adult males in Mixed Adult 1. However, mothers, fathers, and male and female youth are not different by gender in their perceptions of *how often they get to pick what they will watch*.

Adult females consistently more often report that *someone else changes channels when they wish they wouldn't*. The differences are not significant among youth, although the direction is consistent.

Viewing Style by Program Type

In addition to differences in overall viewing styles, females and males also approach particular types of programs differently. (See Table 17.) Adult Cable 1 asked 255 female and 263 male respondents about their viewing of sports, news, movies, and series. Four questions were asked about each type of program:

Table 14.   Channel Changing Motivations

	(from Mixed Adult 2 Study)		
	F	M	p
to turn to a specific channel	2.33	2.15	.365
to see what else is on	1.45	2.00	.003
because of boredom	1.15	1.50	.066
to avoid commercials	0.97	1.28	.130
to watch more than one show	0.41	0.57	.212

Table 15.   Multiple Show Viewing

Study	F	M	p
How often do you watch more than one show at a time?			
Cable Adult 1	24%	42%	.004
Cable Adult 2	1.90	2.07	.000
Playboy	2.71	2.51	.080
Mixed Parent 1	1.85	2.15	.000
Mixed Parent 2	1.83	2.05	.000
5th Grade	2.54	2.66	.011
10th Grade 1	2.40	2.36	.195
10th Grade 2	2.58	2.60	.592

1. How important are _____ on television to you?
2. How often do you plan ahead to watch _____ on TV?
3. When watching _____ , how often do you change channels at commercials and then come back to the show?
4. How often do you change channels during _____ , not at commercials, and then change back to the show?

Both sexes rated news shows as most important to them. Movies were second for females, followed by series and sports. Sports were second for males, followed by movies and series. Sports were significantly less important, and series and movies significantly more important to females than to males.

Planning ahead to watch a particular type of show varies by sex. Males were significantly more likely to plan ahead to watch sports, and females significantly more likely to plan ahead to watch regular series, consistent with differences in the importance of these types of shows. Movie and news planning are not

Table 16.   Control

Study	F	M	p	F	M	p
	How much of the time do you pick what shows you will watch?			How often does someone else change channels when you wish they wouldn't?		
Mixed Adult 1[1]	1.73	2.12	.000	1.08	0.82	.033
Mixed Parent 1[2]	2.35	2.33	.851	1.98	1.36	.000
Mixed Parent 2[2]	2.55	2.53	.798	1.66	1.41	.022
5th Grade[3]	1.93	2.05	.254	2.31	2.15	.175
10th Grade[1]	2.30	2.38	.405	1.89	1.76	.264

For the first item, questions varied by study as follows:
[1]How much of the time do you control what you watch?
[2]How often do you pick what to watch when watching with your child?
[3]How often do you pick what to watch when watching with a parent?

## CHAPTER 13

Channel Types and Viewing Styles

Carrie Heeter and Thomas F. Baldwin

Channels which specialize in a particular genre of content are now widely available over cable television. Program-type preferences have been poor predictors of viewership of broadcast network television. The lack of relationship was typically blamed on broadcast-network scheduling: the mix of program types available changed every half hour, moving from channel to channel. With cable it is possible to examine viewing behavior associated with particular content-specialized channels, removing the earlier obstacle of variable network schedules. This chapter will search for viewing-style patterns across different types of channels. Viewing styles for the following channel types will be contrasted: all-sports, all-news, music videos, pay movie channels, broadcast networks, and superstations.

There is evidence to suggest that different types of content will be watched in different ways. Although program-type preferences do not effectively predict viewing, respondents do seem to have definite likes and dislikes related to program type. Viewing choices may also be linked with motives for watching television at a particular time.

Uses and gratification research has identified different and often contradictory motives people cite for watching television (e.g., to escape, to learn what's happening in the world, etc.: Blumler, 1969). Different people watch for different reasons at different times.

Zillmann and Bryant (1985) have demonstrated that affective state influences choice of program types. An annoyed individual is more likely to select absorbing content that is devoid of provocation, anger, or retaliation, to help disrupt the individual's preoccupation with the experience of annoyance.

Rubin (1984) examined interrelationships between viewing motives and viewing of different types of programs. He characterized two viewing modes: ritualized and instrumental television viewing. *Ritualized* television use is habitual and frequent, encompassing motives such as watching TV for companionship, out of habit, to pass time, etc. That mode is associated with watching action/adventure, comedy, drama, game, and music/variety shows. *Instrumental* viewing is selective and purposeful, relates to information-seeking motives, and is associated with watching news, documentary, and talk/interview programs. Instrumental viewing of news channels should be associated with more planning ahead to watch a particular channel and less channel changing once the channel is initially selected.

Structural programming factors may also influence viewing styles for particular types of channels. For example, program length varies widely, from short music videos to half-hour cyclic news to full length movies to endless sporting events. Structural programming factors may affect viewing style; for example:

- Short cycles like MTV should encourage frequent channel changing and discourage preplanning
- Movies may be best watched from start to finish, requiring more planning. On cable movie channels, there are currently no commercials. Therefore, there is no natural break at which to change channels. The content is often repeated on alternate nights, which may encourage sampling as a precursor to planned viewing.
- Sports events have frequent commercials and very predictable plots, encouraging channel switching. Planning to watch is often necessary, to determine which teams are playing at what time, etc.
- Of the channel types discussed here, on the cable systems studied, the 24-hour cable news channel has the most direct competition when network channels also carry news. In the morning, noon, dinner time, and late evening, there are many channels carrying news shows about essentially the same events. At other times in the day, the cable news channel has no competition. Viewing patterns are likely to be different (longer viewing stretches, less channel changing) during noncompetitive programming periods.

In this chapter, data from various studies reported elsewhere in the book will be re-analyzed to examine viewing style as it relates to specialized channels.

Broadcast network channels will be used as a baseline for "normal" viewership patterns, with diverse rather than specialized content. Superstations will be used as a possible baseline for "normal" cable-only channel viewership patterns.

METHODS

Tables 1 and 2 present self-report survey data on viewing style for movies, news, sports, and series from the studies described in detail in Chapters 10 and 20. Table 1 is based on telephone surveys with 415 cable subscribers drawn randomly from lists of basic cable, one-pay cable, and two-pay cable households, such that respondents were approximately evenly divided across those three groups. Two hundred and one noncable households were also interviewed. Table

2 is based on telephone surveys with 1036 cable subscribers, also evenly split by basic, one-pay, and two-pay. Two surveys were conducted 3 months apart, using the same questionnaire. Half the sample answered questions about movies, news, sports, and series in each wave of data collection.

Table 3 is based on the news viewership analysis of interactive cable data collected over 3 weeks in a new fall season from 164 households in that same community. This study is reported in detail in Chapter 14.

Table 4 shows an analysis of two-way measured channel changes per hour across Weeks 1,2, and 4–5 of a new fall season, by channel type.

Tables 5 and 6 are part of a movie channel viewership analysis which has not been reported elsewhere. Data for Table 6 are the same summer, interactive viewing data used in Chapter 4. Data for Table 5 are the same new fall season interactive viewing data used in Chapter 14. Movie channels were found to be the channels most frequently changed to and away from. Table 6 shows which channels viewers tune *to* when they changed *away from* HBO (Home Box Office) and TMC (The Movie Channel), by channel type.

Table 5 describes how HBO movies are watched, based on interactive household measurement. Eleven movies were identified which were shown at least twice across the sample of available fall data. For every showing of the movies, the total amount of time each household watched was recorded. Viewing periods were broken into five groups: watching the movie for 1 to 10 minutes, 11 to 30 minutes, 31 to 60 minutes, 61 to 90 minutes, and more than 90 minutes. The percentage of movie viewing in each duration category was calculated. In addition, the number of different times each household tuned in to HBO during a showing of the movie was noted, and the average number of tune-ins per viewing

Table 1. Watching Different Program Types*

	Movies	News	Sports	Series
How often do you watch _____?[1]	2.0	2.9	1.8	1.9
How important is it to watch _____ from start to finish?[2]	3.2	2.8	2.6	2.9
How do you usually find out about _____ shows you want to watch?	249	156	231	188
TV Guide[3]	46%	26%	29%	46%
Newspaper TV listing	28%	18%	32%	30%
(Newspaper Sports section)			11%	
Family/friends	8%	10%	15%	9%
Check channels	17%	47%	13%	15%

[1]response categories were: 4 = very often, 3 = quite often, 2 = often, 1 = not very often
[2]response categories were: 4 = very important, 3 = important, 2 = not very important, 1 = not important at all
[3]multiple responses were accepted
*Based on telephone surveys with 415 cable and 201 noncable residents. Data collection was partially funded by Continental Cablevision. Half of the sample answered questions about movies and news. The other half answered questions about sports and series.

Table 2.   Planning to Watch Different Program Types*

	Sports	News	Movies	Series
How important are _____ on television to you?[1]	2.2	2.7	2.5	1.7
How often do you plan ahead to watch _____ on TV?[2]	1.8	1.8	2.0	1.5
When watching _____, how often do you change channels at commercials, then come back to the show?	1.4	1.0	1.0	0.9
How often do you change channels during _____, not at commercials, and then change back to it?	0.9	0.7	0.7	0.7

[1]response categories were: 4 = very important, 3 = quite important, 2 = important, 1 = not very important, 0 = not important at all
[2]response categories were: 4 = very often, 3 = quite often, 2 = often, 1 = not very often, 0 = not at all
*Based on telephone surveys with 1036 cable residents. Data collection was partially funded by Continental Cablevision. Only half of the sample answered questions about each program type.

households was calculated. This in-depth examination of how 11 movies are watched provides a fairly consistent picture of cable movie viewing.

RESULTS

Table 1 suggests that viewers do consciously differentiate to some extent the way they watch different kinds of programming. News is reported to be the most frequently viewed program type of those asked about, even though it is the least planned.

Of movies, news, sports, and weekly series, movies are considered slightly more important to watch from start to finish. However, all four are rated moderately important on the four-point scale. Movie information, more than any other, is sought from *TV Guide*, although the newspaper listing and channel checking are also used. Weekly series and sports information is also derived mostly from TV and newspaper guides. Sports is the program type whose audience is most strongly affected by family and friends' suggestions of what to watch, although the level of impact is not high. News, as previously mentioned, is most often encountered when checking channels, rather than looking it up in a guide of some kind. Presumably, since news over the networks is regularly scheduled, most news program information is derived from general knowledge of broadcast schedules, without need for program listings.

Table 2 shows respondents' differentiation of sports, news, movies, and series in terms of the importance of watching them, planning ahead to watch

Table 3.   News Viewership and Channel Changing by Daypart*

	6–9	9–12	12–1	1–5	5–730	8–11	11–12	12–6
% off-air viewing	65%	56%	59%	58%	64%	66%	59%	48%
% news viewing	54%	8%	39%	15%	39%	6%	43%	7%
% of total news viewing time	20%	4%	8%	14%	34%	7%	11%	2%
% of cable news viewing time	14%	15%	5%	20%	14%	21%	5%	7%
avg channel changes	1.9	2.6	1.4	5.4	5.6	8.0	2.4	2.2
(among viewers)	(4.7	6.0	4.3	8.6	7.4	10.2	4.7	7.7)
avg different channels	1.1	1.3	0.9	2.6	3.0	3.7	1.4	1.0
(among viewers)	(2.7	3.0	2.6	4.2	3.9	4.7	2.7	3.5)

*Data are derived from a content analysis of program schedules for 35 cable channels to determine all news programming and subsequent computation of viewership across a sample of five October 1982 weekdays. Sample size varies by daypart, due to missing data, etc. See News and the Cable Subscriber for a more detailed explanation of methods.

them, and changing channels while the program type is on. News, followed by movies, is the most important content, and series the least important. Viewers plan ahead to watch movies moreso than series. Channel changing is most predominant when watching sports—particularly zapping of commercials, then returning to the sports event in progress. Changing other than at commercials, then returning to the program content, is less common.

Table 3 indicates that, even among cable subscribers with 24-hour news services available, news viewing is much heavier in the periods when broadcast stations are presenting news: 6 to 9 a.m., noon to 1 p.m., 5 to 7:30 p.m., and 11p.m. to midnight. Conversely, cable news viewing is most heavy in the dayparts when broadcasters have little or no news, particularly 1 to 5 p.m. and 8 to 11 p.m. Newsviewing habits established by broadcasters' schedules over the years have not been affected significantly by the massive availability of cable information. On the other hand, some cable subscribers are taking advantage of

Table 4.   Channel Changes Per Hour by Channel Type* (Prime Time Trend Analysis)

Channel Type	Week 1	Week 2	Week 4–5
Network	4.2	3.7	2.9
Other Off-Air	5.6	7.1	6.5
Pay Movie	8.4	6.2	5.9
Other Cable	10.3	10.9	9.3
Overall	6.4	5.9	4.9

*Data are derived from analyses of eight prime-time weekdays in September and October 1982. (See Chapter 6.)

Table 5. Movie Channel Viewership: Duration of Viewing

Movie Title	Date	Time	Viewing Households by Time					Total HH	Tuneins	Avg.
			1–10	11–30	31–60	61–90	90+			
Adventures of the Wilderness Family	10/21	8 am/3 pm	17	4	5	1	2	29	54	1.9
	10/24	12:30 pm	25	3	1	3	2	34	56	1.6
Arthur	10/1	8 pm	16	4	9	2	1	32	60	1.9
	10/22	10 am/8 pm	24	14	2	4	8	52	149	2.9
Camelot	9/29	8 pm	26	1	5			32	57	1.8
	10/8	9:30 pm	21	2	4	1		28	50	1.8
1st Monday in October	10/6	8:30 pm	21	3	6	4	6	40	82	2.1
	10/21	11:45 pm	16	3	1	1		21	37	1.8
Green Ice	10/11	8 pm	17	6				23	25	1.1
House Calls	10/23	8:30 am/6 pm	41	14	14	4	6	79	147	1.9
	10/23	2 pm/11 pm	14	12	11	6	2	45	79	1.8
	10/25	8 pm	20	4	3	3	3	33	48	1.5
Islands in the Stream	10/5	8 pm	13	1	1			15	24	1.6
	10/21	1 pm/8:30 pm	30	7	2	2	4	45	87	1.9
Money Matters	10/22	4:30 pm	5	4				9	10	1.1
	10/26	7:30 pm	12	6				18	22	1.2
One on One	10/24	8 am	26	10	4	7	6	53	145	2.7
	10/27	8 pm	10	3	1	3	9	26	58	2.2
Silence of the North	10/24	11 am	12	8	4	3		27	52	1.9
	10/26	11 pm	4	1	3		1	9	9	1.0
Summer Solstace	10/21	10 am	2		1			3	3	1.0
	10/24	2:30 pm	7	2	5			14	16	1.1
TOTAL			422	120	85	46	54	727	1381	1.9
Percentage			58%	17%	12%	6%	7%	100%		

Table 6.   From Movie Channels to . . . A Zapping Analysis

Channel Changed to:	HBO Tuneouts		TMC Tuneouts		
	(n)	% of tuneouts	(n)	% of tuneouts	Average
Movie Channels					21%
TMC	88	23%	—	—	
HBO	—	—	90	18%	
MTV (Music)	51	13%	64	13%	13%
Networks					8%
NBC	32	8%	43	8%	
CBS	38	10%	36	7%	
ABC	18	5%	45	9%	
ESPN (Sports)	19	5%	21	4%	5%
Superstations					4%
WTBS	17	4%	30	6%	
WOR	21	5%	22	4%	
WGN	11	3%	18	4%	
CNN (News)	8	2%	14	3%	3%
20 Others	63	18%	101	21%	19%
Total	371		484		

their 24-hour news availability and concentrating their cable-news viewing in the hours when broadcasters do not supply it.

An average of 1 hour and 21 minutes per day was spent watching "information programming," and 20 minutes of that was on cable-only channels. The news viewing time by daypart was highly consistent across the 5 sample days.

It appears that entertainment programming on cable may be attracting cable viewers away from news. In periods where people who do not have cable have little choice but to watch broadcast news, that is, in periods where the major broadcast stations are only offering news, less than half of the cable subscribers are watching news.

Still looking at dayparts, the bottom of Table 3 shows different levels of channel-changing activity at different times in the day. Whether this is due exclusively to a different composition of viewers, or may in part be due to different modes of watching by the same viewers, remains to be determined. Using households which viewed at all in each daypart as the divisor, more channel changes occur per hour between 1 p.m. and 11 p.m. and after midnight, ranging from seven to ten changes an hour, with the largest number of changes occurring during prime time. Morning channel changing was less frequent. Similarly, more different channels were watched in those same time slots (3.5 to 5 different channels). Viewing is quite active, and the activity level varies by daypart.

Table 4 suggests that the activity level also varies by network season. Overall channel changes per hour decrease from 6.4 in Week 1 of the new season, to 5.9 in Week 2, to 4.9 in Weeks 4–5. More important for the purpose at hand are the channel type comparisons.

Commercial network channels are watched with the fewest channel changes per hour. Pay movie channels receive considerably more changes per hour, and other cable channels are sampled even more frequently (nine to eleven times an hour).

Turning to a more in-depth examination of movie viewing, Table 5 reports the duration of watching 11 movies broadcast two or more times on HBO. Of the 727 separate viewing times identified, 58% of the tune-ins stayed with the movie for 10 minutes or less. Another 17% lasted half an hour or less; another 12%, 1 hour or less. In sum, at least 87% of the time a household watched at least part of a movie, the household watched less than the entire movie, and generally a lot less. This is not necessarily to conclude that households never watched whole movies—the HBO movies are broadcast many different times, and, within the sample of 162 households, it may not be unreasonable that only 7% watched a whole movie across the 27 showings when data were collected. It is, however, to conclude that households often watch parts of movies. Those households which did tune in to HBO while a movie was being broadcast tuned in an average of twice during that broadcast. The period of watching the movie was the sum of all tune-ins during that broadcast, but number of tune-ins was tabulated and averaged. An alternative interpretation may be that viewers watch whole movies, but across multiple showings. The level of repeat viewing (of movies already seen) may also be more likely to occur in the form of repeat viewing of parts rather than all of a movie.

One final characterization of movie-channel viewing can be added. In a two-pay channel cable system, watching (or checking) one pay movie channel was most often followed by watching (or checking) the other pay movie channel (Table 5). (Pay-channel penetration in this cable system was 165%—a majority of cable subscribers subscribed to both pay services.) Twenty-three percent of all tune-outs from HBO went to The Movie Channel; 18% of all tune-outs from TMC went to HBO.

This suggests that viewers approach the movie channels with the same viewing style, by channel type. The second most popular tune-out destination for both TMC and HBO was MTV (13%), followed by the commercial networks at 10% or lower.

CONCLUSIONS

The major empirical findings about how pay movie channels are watched are summarized below.

1. Viewers indicate that movies are slightly more important to watch in their entirety than other program types.
2. However, much more viewing of parts of movies occurs than does watching them from start to finish.

a.   Of those who watch any of an HBO movie when it is broadcast, about 7% may actually watch the whole movie. This is quite consistent, regardless of the movie's popularity or time of day it is broadcast. More than half of viewers tune in for 10 minutes or less.

b.   Twenty percent of tune-ins to pay-movie channels result in viewing stretches of 15 minutes or more.

3.   TV guide, newspaper listings, and channel checking are all used to seek program information. Among households which do tune in to an HBO movie, that movie is turned to an average of two different times during that broadcast.

4.   There is a distinctive pattern of channel checking, suggesting that movie channels are watched in a consistent manner different from watching the movie from start to finish.

a.   The most likely place to turn to after checking HBO is TMC, and vice versa. This suggests a "movie-checking" mode.

b.   Second most likely for both movie channels is MTV, a channel with short segments. With its brief 4- to 5-minute segments, MTV seems a more likely candidate for watching in short periods than the movie channels. However, the movie channels appear to be used in similar ways. Perhaps movie channels, in addition to being watched for movies from start to finish, may also serve a similar function to MTV's, providing variable length content which doesn't have to be watched as an entire program segment.

Here are the major empirical findings for news viewing:

1.   News is reportedly the most frequently/regularly viewed program type. If planned, the planning is based more on memory of broadcast schedules than on program guides.

2.   Channel checking is the most important information source about news programming.

3.   News comprises 40% or more of all viewing 6–9 a.m., noon–1 p.m., 5–7:30 p.m., and 11 p.m.–midnight (periods when the networks offer news). More viewing of news from cable occurs in periods when networks do not offer news.

4.   Like the movie channels, 20% of CNN viewing results in viewing stretches of 15 minutes or more, accounting for more than three-fourths of CNN viewing time. These channels are sampled much more frequently than they are watched.

   The data that are brought together in this report do not completely clarify how news and movies are watched. But they do shed some light. Various factors are confounded within this analysis. First, the analysis for movies is only of movie viewing on 24-hour movie channels. The picture that emerges is quite clear—

that partial movie viewing is the norm, and the movie channel is one of several checked in sequence. But whether this is typical of movie viewing, or unique to movie viewing on pay 24-hour channels, has yet to be tested. The viewing is active in the sense that viewers tune in (and out) an average of twice within a given movie broadcast. It is likely that the viewer is exercising selectivity in what they watch on those channels—perhaps engaging more in instrumental than ritualistic viewing.

News viewing, if it is an example of instrumental viewing, is curiously not something viewers seek in a guide. The news channel is checked more frequently than it is watched, in the same proportion as the movie channels. The concentration of news viewership in periods of network news broadcast suggests that either viewers would watch something else if it were available, or their news viewing is quite habitual and they maintain the network schedule despite other available times.

Cable channel viewing may be more selective than network viewing. The channels are checked more often and watched less often. Perhaps, when they are watched, it is truly content the viewer is interested in.

Program Type and Viewing Style

## CHAPTER 14

News Viewing Elaborated

Thomas F. Baldwin, Carrie Heeter, Kwadwo Anokwa, and Cynthia Stanley

Cable subscribers may have access to more than 20 television channels with news and information programming. Unlike broadcast television viewers, they are not forced to view news at fixed times, or to view public affairs programs only at the odd hours that are undesirable for entertainment programming. The cable subscriber has news channels available around the clock. Some people in small towns or suburbs have television news of their own communities for the first time through cable-local origination and access channels.

How much of this news and information do cable subscribers watch? Do they take advantage of the 24-hour availability of news? Does cable news displace broadcast news in cable homes? Is cable news used during periods when broadcast outlets are offering only entertainment programs? What proportion of news and information viewing occurs on channels available only over cable? These are the principal questions of this study. The data come from minute-by-minute monitoring reports of a two-way cable system.

BACKGROUND

Broadcast television news has often been criticized for its news scheduling; early morning, noon, early evening, and late evening. These times may not be convenient for some viewers on a given day, and indeed may not be available at all to some households (i.e., the home with parents of young children too busy for the news in the early evening, and too exhausted in the late evening). In-depth public affairs programs on broadcast television are relegated to Sunday mornings and late, late evenings. Broadcast news is not prime-time programming. Furthermore, broadcast television time is too valuable for continuous live coverage of breaking news.

Pioneering cable entrepreneur Ted Turner addressed these problems with the Cable News Network (CNN), the first 24-hour television news service. It was followed by the Westinghouse-ABC 24-hour service, Satellite News Channel (SNC). Turner countered with his own CNN Headline News to compete, and later bought the SNC. Most of the SNC subscribers have now been absorbed into the CNN or CNN Headline networks, which, at the beginning of 1985, had 30,564,000 and 13,584,000 subscribing households, respectively. Other 24-hour news and information services include the Cable-Satellite Public Affairs Network (C-SPAN, 20,000,000 households), which includes a gavel-to-gavel coverage of the U.S. House of Representatives, panels, interviews, and speeches; The Financial News Network (17,500,000 households), with business news; and The Weather (15,700,000 households). Automated television text channels are also available in many cable households, e.g., AP News Cable, Dow Jones Cable News, and Reuters News View. Other text channels in some systems include school- and government-news bulletins and announcements, shopping channels with marketbasket prices, and community events calendars.

Cable networks such as Black Entertainment Television, Arts and Entertainment, ESPN, Lifetime, Modern Satellite Network, Satellite Program Network, Spanish International Network, and Superstations WGN, WOR, WPIX, and WTBS all carry news and information as part of a broader programming base. Home Box Office and Showtime, pay networks, carry occasional documentaries and consumer-information programs. Even the Playboy channel has a somewhat specialized, regularly scheduled news program.

Add to this all the televised meetings, news and public affairs programming on government, educational and public access channels, and the cable-company-programmed local origination channels.

A high capacity, full service cable system may have five full-time news networks, several full-time text information channels, at least three access channels, at least one local origination channel, and 13 cable networks, all with regularly scheduled information programs.

From the standpoint of television economics, reality programming is generally cheaper than the creation of fiction or entertainment programming. Information is necessary to fill the single or dual cable systems that may supply more than 100 channels.

Theoretically, there is great value in the abundance of news on cable. It is always accessible, at the convenience of the subscriber. Full live coverage of major local, regional, national, and international events is available. Communities once swallowed up in metro television markets now have a television identity. Channel explorers with only a few minutes of time for television can phase into news programming more easily than try to comprehend the plot of an entertainment program in progress. Commercial zappers may briefly tap into news channels for the same reason.

Whether cable subscribers avail themselves of these opportunities is the question here. Little is known of the cable audience for news. Only CNN, of the cable news services, has been regularly surveyed by commercial ratings services. It manages to earn a rating point or two against its broadcast and cable competition.

AUDIENCE RESEARCH

There are only a few studies relevant to the main concerns of the present research. Reagan (1982) examined the impact of cable on the use of news in general and on the traditional uses of local and network broadcast news. The study was based on telephone interviews with cable subscribers and nonsubscribers in Ann Arbor and Grand Rapids, Michigan. He found that cable subscribers spend equal or more time with network and local television news than do nonsubscribers. CNN was used by subscribers but was viewed less than were the other news media, including network television. Generally, CNN was considered as believable and useful as the other media. The study found that CNN was used throughout the day and that CNN viewers spent considerably less time with CNN than with other news media. From this, Reagan suggests CNN is used as a supplemental service, perhaps a filler as the viewer switches channels to CNN when there is a break or dull moment in other programming.

A study by Henke, Donohue, Cook, and Cheung (1983) also examined the effects of cable on television news viewing habits and preferences. The research, conducted in Lexington, Kentucky, was based on telephone survey of cable households. It was found that cable subscribers watch less broadcast network news than nonsubscribers, and the same amount of local news. The researchers state that there were CNN viewers among the cable subscribers (the number increasing with length of time as a subscriber). They suggest that CNN may substitute for broadcast network viewing, since it is a national-international service, but, because of the lack of local coverage, CNN does not affect local news viewing. They also conclude that precable news viewing habits "die hard," given the relatively slow pick-up of CNN.

A commercial firm, Audience Research and Development, using a national sample, also found that CNN and SNC did not significantly affected local television news audiences (Gelman, 1983). Only 1% of the cable subscribers claim that these cable news sources changed their local-news viewing habits. The research company stated that viewers "apparently do not see existing cable news options as a replacement for local newscasts, but instead as a supplement."

There were some sidelights to this major conclusion, however. Subscribers who said they watched less news than they used to were pay or multipay subscribers. About 40% of the cable subscribers confessed to occasionally skipping the news for entertainment. Heavy viewers of cable news were also heavy viewers of local broadcast news. There was a correlation between heavy viewing of ESPN (cable sports network) and reduced viewing of local news. Perhaps the sports newscasts on ESPN eliminated that motivation for watching local news.

Kaplan (1978), surveying cable subscribers in Woodland, California, by phone, found that respondents claimed to use automated information services with some frequency: "47% of the respondents used at least one type of automated information channel (weather, sports, news, community information and financial) four or more times a week" (p. 158).

Research by Jeffres (1978), on the effects of the introduction of cable in a small community raises some interesting hypotheses about news viewing. Jeffres expected that people of low socioeconomic status would watch less news, and that high-SES persons would watch more news, when they began to receive cable. Jeffres queried households about viewership before and after the introduction of cable (in late spring). Overall viewership of television, and news viewership, dropped after cable was installed, presumably a function of the normal seasonal flux in viewership. High-SES persons were found to watch less local news than low-SES people after cable. This led him to speculate that high-SES people increased their proportion of viewing time given to national and international news, diverting from local news.

If these assumptions were to be verified, we could see cable changing the information exposure among higher- and lower-SES groups along localite–cosmopolite dimensions. Cable could make television more like newspapers and magazines in the ability to discriminate among viewers, in this dimension.

Other research on cable has addressed the matter of diversion of the television news to distant city local news. A study was conducted in Bryan College Station, Texas, a community with two local television stations. Most households in the community were able to access another eight broadcast stations from distant towns, only one of which could be received off air. These cable subscribers were surveyed by telephone. Thirty percent of the local news viewers were diverters to distant local television news. While it is possible *some* television viewing areas to watch *broadcast* news from distant stations, almost all cable households have this capability.

Grotta and Newsom (1982) also document the diversion of the cable audience from local broadcast stations but note an apparent relationship between cable subscription and perceived need for information. Cable subscribers, they found, spend more time with local daily newspapers and radio. Among cable subscribers, other local media may make up for the snubbing of local TV.

All of these studies rely on recall of channel use (difficult in a multichannel environment) and, in some cases, recall of before-cable viewing. In this study, we will use records of actual household set use.

DATA COLLECTION

Data were collected in a two-way cable system in Temple Terrace, Florida, a suburb of Tampa-St. Petersburg, for one full week, October 21 through October 27, 1982, from a randomly selected sample of 200 households. Replacement with adjacent households was permitted where necessary. The headend computer in the cable system scanned viewing behavior in all 200 households every 20 seconds, storing the *time* of each change in status (either on/off, or changes to different channels), the *channel changed to*, and the *household identification number* whenever a change occurred. The data were stored on floppy disc and sent to the Communication Technology Laboratory at Michigan State University by Coaxial Communications, owner and operator of the cable system at the time of the study. From there, data were transferred to the Cyber mainframe computer at MSU for analysis. The data were recorded on to floppy disc whenever the memory buffer filled up, resulting in 10–15 data files per day. System-wide errors and problems in data transfer resulted in the loss of individual data files. One sample day had complete, usable data, and the others had varying portions of the day missing. In order to take maximum advantage of available data, each weekday was divided into eight dayparts, reflecting different periods of news availability. Periods of broadcast news availability included 6–9 a.m., 12–1 p.m., 5–7:30 p.m., and 11 p.m.–12 midnight. Periods of little or no broadcast news were 9 a.m.–12 noon, 1–5 p.m., 8–11 p.m. (prime time), and 12 midnight–6 a.m. Table 1 lists the dayparts for which data exist, for each sample weekday. The period 7:30 p.m.–8 p.m. was not analyzed, as it was deemed more important to examine prime time (8–11 p.m.) and the period of broadcast evening news in the market (5–7:30 p.m.) as complete units.

The viewership data reported in this study represent an average weekday, and were created by conducting separate analyses of each daypart in each day and averaging across days where data existed. Thus, data reported for the 6–9 a.m. period reflect average viewership across Tuesday, Thursday, and Friday, while 11 p.m.–12 midnight average Monday and Thursday.

Across the sample 200 households, 36 of the terminals registered errors for 50% or more of the data collection period. They were removed from all analyses. Among the 164 households retained for analysis, the error rates were very low, ranging from 1%–5% averaged across all households for any daypart, and never exceeding 49% of the daypart for an individual household.

In addition to the monitored viewing data, the households were surveyed. Mail questionnaire was used, with overt household identification. Those households which did not return a questionnaire within 10 days were interviewed by phone. Surveying occurred approximately 6 months after collection of viewing

Table 1. Data Availability by Day Part

		6–9 a.m.	9–12 noon	12–1 p.m.	1–5 p.m.	5–7:30 p.m.	8–11 p.m.	11–12 p.m.	12–6 a.m.
Monday	10/25					X	X	X	
Tuesday	10/26	X	X	X		X			
Wednesday	10/27			X					
Thursday	10/28	X	X	X	X	X	X	X	X
Friday	10/29	X	X	X	X	X	X		

data; thus, respondents were not sensitized when the system was being used for data collection. One hundred and fifty of the 200 sample homes completed surveys; 39% were by telephone, 61% by mail. The average age of the adult respondent was 47. The mean education was 2 years of college. Average household income for the cable subscribers fell between $30,000 and $40,000 (within cable subscribers, which may be expected to be slightly higher than the overall population). Households contained an average of 2.3 adults and .6 children. While these demographics suggest our data may not be generalizable to the entire population of cable viewers, given upscale income and educational levels, they do provide an opportunity for the initial examination of the use of cable for reality programming.

DESCRIPTION OF CABLE SYSTEM

The Florida cable system contained 35 channels, including two pay services (HBO and The Movie Channel), four network stations, four independents, three superstations, two PBS stations, four local access channels, and 16 other cable-only channels. There were two 24-hour all news channels, CNN and SNC, in addition to several automated channels with continuous informational content: Reuters Stock Service, a weather channel, a local newspaper channel with news and classified information, a local advertising channel, plus bulletins on the government access, educational access, public access, and local origination access channels.

AVAILABILITY OF INFORMATION BY CABLE

There are 12,634 minutes of information available on the typical *weekday* on the sample cable system—211 hours. Nearly 5,000 of these minutes are in the general news category, and over 6,000 minutes are topical news (business, sports, or other specialized information). With cable, news availability time increased over that available in the community from locally available broadcast channels by a whopping 685%; general news increased by 482% and topical news by 5,129%. Table 2 presents the results. Importantly, cable adds 5,428 minutes of news and information in periods when there is *none* available to the noncable household.

**Table 2.   Minutes of News Availability
on a Single Sample Weekday**

News Category	Off-Air Channels	Cable-Only Channels
General news	1,020	4,719
Topical news	120	6,155
Other	470	150
Total	1,610	11,024

NEWS PROGRAM CONTENT AND VOLUME

The news content was separated into local broadcast and cable-only categories. The broadcast category included the programming from local broadcast stations that would have been available to our sample of cable subscribers even if they were not on cable. The cable-only programming included all of the local-originated, cable network, and text channels available only to cable subscribers, as well as broadcast stations that were imported from distant cities.

News programming was categorized as: *general*, typical TV package of news, sports, and weather; *mixed*, news and feature information news; *magazine*, package of short informational segments; and, *other*, public affairs, documentaries, news interview shows.

Table 3 presents the categories and the amount of time available in each category for basic cable subscribers on a typical weekday. It demonstrates the tremendous volume of news and information available to subscribers to the system. In this case, the cable subscribers had both CNN and Satellite News Channel. Now, typically, such a system would have CNN and CNN Headline News, so that there would be no diminution of news volume since the demise of SNC.

Cable introduces the category of "topical" news; the sports news shows, the business reports, weather, etc. Exclusive programs of a half-hour or hour in duration about a particular topic, let alone whole channels devoted to a single topic, are simply not available to broadcast audiences, with the exception of a few PBS programs.

Table 3 was used to develop columns 4 and 5 of Table 4, the aggregate hours of nontext cable news and broadcast news available to cable subscribers. Basic cable subscribers have 22 hours of broadcast news from their own broadcast market and another 61.5 hours of nontext news supplied by cable channels. The cable subscribers who opt for pay channels add a few more hours per week of documentaries and information programs on those channels. The 9 a.m.–12 noon, 1–5 p.m., 8–11 p.m., and 12 midnight–6 a.m. dayparts are periods when cable is the primary source of available news programming.

USE OF INFORMATION BY CABLE SUBSCRIBERS

The actual use of all this information by cable subscribers is another matter. On a typical weekday the average cable household in our total sample viewed television 6 hours, 15 minutes, slightly under the national average for all television households. Of this time, a substantial 1 hour, 21 minutes was given to information programming (22% of total viewing time) and 21 minutes of the information viewing was on cable-only channels (25% of the total information viewing time, which includes broadcast news viewing).

Cable-only information viewing was evenly split between general and topical news (53% general). Nine minutes of the cable information viewing of the aver-

Table 3. Amount and Type of News and Information Available in Temple Terrace by Day Part on Weekdays (*Cont.*)

Day part*	Hrs. available ( ) = text channels	News type b = broadcast c = cable	Examples of news type

6–9 a.m.	BROADCAST
	General 3.0
	Mixed..................4.0
	
	CABLE ONLY
	General 6.0 (3)
	Mixed..................2.0
	Topical.................0.5 (12)
			Network news, PBS weather
			CNN, SNC, Tampa News
			Today, Good Morning America
			Good Morning America
			ESPN's Sportscenter, govt.,
			    Reuters stocks, shopping,
			    weather

9 a.m.–12 noon			INN from WTOG
			News from WTBS, INN from
			    WOFL, CNN, SNC, Tampa
			    News
			Variety news from SIN
			Public affairs, talk-discussion
			    shows
			Govt., Reuters stocks, weather,
			    shopping

12 noon–1 p.m.			Local news
			News from WOR, CNN, SNC,
			    Tampa News
			Pulse Plus (local)
			Reuters stocks, weather,
			    shopping, govt.

1–5 p.m.	8.5 (4)
	
	.5 (16)
	
	1
		c-general
		
		c-topical
		
		b-magazine
			INN on WGN, CNN, SNC,
			    Tampa News
			Sportscenter on ESPN, weather,
			    Reuters stocks, shopping, govt.
			Hour Magazine

5–7:30 p.m.	4.5
	7 (2.5)
	
	1
	.5
	.5
	
	(10)
		b-general
		c-general
		
		b-topical
		b-magazine
		c-magazine
		
		c-topical
			3 hours local, 1.5 hours network
			1.5 on ABC affiliate, .5 on SIN,
			    CNN, SNC, Tampa News
			Nightly Business Report on PBS
			PM Magazine
			People and Places (Local
			    origination)
			Weather, Reuters stocks,
			    shopping, govt.

8–11 p.m.	1
	8.5 (3)
	
	1
	.5
	
	.5 (12)
	
	.5
		b-general
		c-general
		
		c-magazine
		b-topical
		
		c-topical
		
		b-other
			Local news on WTOG
			News on WTBS, WGN, SIN,
			    CNN, SNC, Tampa News
			20/20
			MacNeil/Lehrer, Sneak Previews
			    on PBS
			Sportscenter on ESPN, weather,
			    shopping, govt., Reuters stocks
			MacNeil/Lehrer

Table 3.   (*Continued*)

Day part*	Hrs. available ( ) = text channels	News type b = broadcast c = cable	Examples of news type
11 p.m.–12 mid	2.5	b-general	2 hours local, .5 hours network
	3.5 (1)	c-general	INN, WXLT, CNN, SNC, Tampa News
	.5 (4)	c-topical	Sportscenter on ESPN, weather, Reuters stocks, govt., shopping
12 mid–6 a.m.	1	b-general	Latenight on PBS, local news
	(18)	c-general	CNN, SNC, Tampa News
	.5 (24)	c-topical	Sportscenter on ESPN, govt., Reuters stocks, weather, shopping

*5–7:30 p.m. represents the period when broadcast evening news is available in the market. 8–11 p.m. is prime time. The half-hour not reported (7:30–8 p.m.) included no broadcast news, and 1 hour of nontext cable news.

age household were given to the 24-hour news channels. Looking deeper into this viewing, we discover that absolutely none was given to the automated channels (e.g., Reuters, stocks, local newspaper, shopping channel, weather). This means that all of the 24-hour news channel viewing was with CNN or SNC. Eighty percent of the cable households never watched these channels. Thus, CNN and SNC had no impact on four-fifths of cable subscribers and a very substantial impact on the remaining 20%. Among those who watched CNN or SNC, an average of 45 minutes was spent with the channels.

Pursuing this area further, we found that 48%, just about half, of the cable households did not use the cable-only information services at all. This compares to only 27% who did not use off-air broadcast television information services. Thirteen percent of the nonuser homes on a given day, in both cases, did not view television at all.

Some of the cable people could be considered cable news freaks. Five percent of them viewed more than 2 hours of cable-only information in a day (mainly on CNN). But the availability of cable information is not necessary to heavy information exposure. Fourteen percent of the cable households watched more than 2 hours of off-air broadcast information.

DAY PART ANALYSIS

Table 4 describes the news viewing to cable-only and broadcast channels of television viewing households in the sample. The viewing data are averaged over Monday through Friday of the sample week. (Columns 4 and 5 are based on the content analysis of news programming on a typical weekday.) Only those households actually viewing television during the daypart are included. News viewing habits die hard, as others have observed, despite the scheduling flexibility of-

Table 4. Cable-only and Broadcast News Viewing by Weekday Day Part Among Viewers

Day part	Number of viewing H/H (n = 164)	(1) % of day part viewing time to news	(2) % of day part viewing time to cable	(3) % of day part news viewing to broadcast	(4) Aggregate hours of non-text cable news available in day part	(5) Aggregate hours of broadcast news available in day part	(6) Average minutes of cable news viewed by H/H	(7) Average minutes of broadcast news viewed by H/H
6–9 a.m.	66	54.4	17.1	82.9	8.5	7.0	6.9	34.3
9 a.m.–12 noon	71	7.6	79.6	20.9	9.5	.5	7.0	0.2
12 noon–1 p.m.	65	39.3	15.4	84.1	3.0	2.0	2.6	13.8
1–5 p.m.	104	15.3	34.2	65.8	9.0	1.0	6.2	12.2
5–7:30 p.m.	126	38.6	9.7	90.1	7.0	6.0	3.5	32.0
8–11 p.m.	129	6.0	73.3	26.7	8.5	2.0	5.6	2.0
11 p.m.–12 mid	84	42.5	7.3	92.7	3.5	2.5	1.3	16.1
12 mid–6 a.m.	47	6.9	86.2	13.7	12.5	1.0	4.4	.7

fered by cable. In column 1, it can be seen that time periods in which the greatest proportion of news viewing time occur are those dayparts which are traditional broadcast news periods: 6–9 a.m., 12 noon–1 p.m., 5–7:30 p.m., and 11 p.m.–12 midnight. In these dayparts, most of the cable households are viewing broadcast news channels; between 80% and 90% in each of the four periods as shown in column 3. The fear that cable subscribers are denying themselves local television news does not seem to be justified by these findings. Only about 10% of the cable-household news viewers are being diverted from their local news to news from other cities or cable network news. This assumes, of course, that these cable-only news viewers would view local news if there were no alternative.

Conversely, news viewing on cable channels is greatest in those time periods which are *not* traditionally broadcast news periods and there are relatively few aggregate hours of broadcast news available (columns 2 and 5). Of all the news viewing in the daypart, cable-only channels command a large proportion in the 9 a.m.–12 noon, 8–11 p.m., and 12 midnight–6 a.m. periods. In these relatively long blocks of time, there is very little news and information available from broadcast sources. The only divergence from this pattern is in the 1–5 p.m. daypart, when available broadcast television aggregates only one hour of news but attracts two-thirds of the news viewing time for the whole period, against the aggregated 9 hours of available cable news programming.

DISCUSSION

Cable household viewing of 1 hour and 21 minutes of information, slightly more than one-fifth of the total viewing, seems substantial. The fact that 21 minutes of this is to cable-only channels suggests that cable makes its contribution to news consumption. Since nearly half of the cable households did not use cable information channels at all, those households that did were using almost 40 minutes daily of cable-only information.

But only 20% of the cable households watched the 24-hour news channels. The finding that 80% of the cable subscribers ignored the all-news channels, and that half of the households ignored cable information of any kind, is interesting. Apparently, as Henke et al. (1983) conclude, old news-viewing habits are resistant to change. There is also some evidence that cable subscribers have a limited awareness of what is available. And some subscribers may be aware of the information channels on cable and have rejected them. The industry is attacking this problem through the Council on Cable Information by attempting to enhance the image of cable programming in general through a national advertising campaign, and, through local tie-ins, the specific program services.

The daypart analysis makes clear that some subscribers are taking advantage of the pervasiveness of cable information. Cable news viewing peaks when broadcast stations are not presenting news. In fact, cable does not seem to detract much from broadcast news. Ninety percent of the subscribers stay with local market broadcasting stations during the periods when local news is broadcast. On the other hand, one might worry about the remaining 10% who are viewing cable news options, all distant, and are not focusing on local events. If cable subscribers are only just beginning to be aware of their news and information options through cable, this small minority of viewers, attracted away from broadcast news by nonlocal information programs, may grow.

For our purposes, the monitoring of set use is superior to other means of determining news viewing which rely on recall. Recall is not especially reliable for cable viewing, given the number of options provided by cable and the remote tuners which make channel hopping convenient. Monitoring suffers the defect, however, of presenting only households, rather than individual, data. Because cable offers an opportunity for much more individualized viewing, this is unfortunate. It would be useful to know how this potential affects news consumption for individuals within the household.

## CHAPTER 15

Playboy Viewing Styles

Bradley S. Greenberg, Carrie Heeter, and Carolyn A. Lin

The bulk, if not all, of the viewing-style evidence comes from the general situation rather than the specific. That is, viewers are not observed while watching television, nor while watching alternative types of programming. They are asked to reconstruct, to recall their typical mode of watching television. There are obvious difficulties with such recollections: has the viewer ever thought about the issues raised in the questions asked; does the viewer think about the most recent viewing situation as the most typical; does the viewer think about the most recent activity because that is the extent to which memory is reliable; and so on.

These are all valid questions about the methodologies employed. Yet equally salient is the issue of the variability of an individual's viewing style. In Chapter 8, we explored the question of the stability of viewing-style variables over time, in the context of an intervention with a cable guide. Here, we address the question from the standpoint of the content of what is being viewed. For example, is there reason to suspect differences in viewing styles while watching news programs in contrast with music videos, or situation comedies, or movies, or sports? Each of these is available exclusively on alternative cable channels, as well as being generally available on commercial broadcast networks. Is there more zapping with one of these, or more guide checking, or more advance planning, or more dual-channel watching? We suspect there is.

For example, the typical live sports event on television has considerable "down-time." In the fall of 1986, the normal National Football League television game consumed about 3 hours and 15 minutes of air time. If you chose to watch such a game by video recorder, you could cut your viewing time by more than 60%. You could zip through the pregame commentary, commercials, the time outs, half-time, replays, etc. Actually, you could watch the active parts of the game in about 1 hour and 20 minutes. So, with nearly 2 hours of filler in the live broadcast, the viewer has much time on his or her hands. And perhaps a remote control or cable channel-changing device in one hand. Either permits the viewer to take advantages of the game's lapses to watch elsewhere—a music video, a second football game, sufficient segments of a dramatic program to follow the story, etc. With this kind of content the viewer has more opportunity, and perhaps greater motivation, to maximize television viewing time. We would anticipate increases in commercial zapping, more multiprogram watching, more channel checking to see what else is on, and perhaps some additional planning of that particular viewing, i.e., knowing what games are on and when, among other possible deviations from what is reported as typical.

So what is reported as one's typical viewing-process behavior may represent an average, and this average may vary considerably across different content modes. The present study addresses this issue by examining viewing as it pertains to a single cable channel, and re-addressing it with reference to television watching in general. At least from a self-report, how sensitive are viewers to alternative uses of the viewing behavior tendencies they acknowledge? An opportunity was extended to collect data from a sample of subscribers to the Playboy Channel. In doing so, the members of the subscriber sample were asked to think about their viewing of the limited kinds of content available on that channel—movies, panels, games—and were queried separately about their general viewing behaviors.

METHODS

Each week, the Playboy Channel conducts a subscriber survey in four of its markets, an attempted survey of 100 subscribers in each market. For the purposes of this study, the research director of the Playboy Channel agreed to add questions on viewing style to its standard questionnaire. The markets in which data were gathered were: Memphis, Tennessee; Columbus, Ohio; Groton, Connecticut; and San Diego, California. The obtained sample was 304, evenly split across markets.

The sample group had the following demographic attributes:

- Gender: 74% male
- Ethnicity: 89% White
- Marital status: 65% married, 20% single, and 12% divorced.
- Children: 43% had children under 18 at home
- Age: 27% were 25–34, 41% were 35–49, and 15% were 50–54.
- Education: 39% graduated high school and 20% each had some college or a bachelor's degree.
- Income: 16% were under $10,000, 24% from $20–$30,000, 20% from $30–35,000, 17% from $35–50,000, and 10% over $50,000, with 13% refusing to answer.

Variables. In accord with the viewing model presented in earlier chapters, here we assessed for this singular channel the manner in which viewing was reported to occur. In particular, measures were constructed to assess the *orienting-*

search mechanisms—channel checking, guide use, and forward planning—and *re-evaluation. Channel repertoire* was also assessed, as was *repeat viewing*, a variable developed especially for a channel which does extensive repetition of its programs.

The viewing process reported for this channel by its subscribers will be compared with their general viewing pattern for other channels. In doing this, we will provide descriptive information as to levels of use of process measures, as well as comparisons between the Playboy and non-Playboy viewing situations.

Finally, predictive models for guide use, channel use, and channel satisfaction will be presented. To do this, some descriptive information with regard to use and satisfaction with the Playboy Channel will be included.

RESULTS

For most of the viewing process measures, a five-response category scheme was used. Do you do this: ''Almost all of the time . . . 3/4 of the time . . . 1/2 of the time . . . 1/4 of the time . . . (and) . . . Almost none of the time.'' These responses were scored 1–5 in sequence, so that a lower score reflects more frequent use of a particular behavior.

Playboy vs. Other Viewing

Respondents were asked how they watched television in general and how they watched the Playboy Channel (*PC*) in particular. The basic comparisons are contained in Table 1. The text provides additional information.

Orienting search and planning.   This phase of viewing reflects a number of potential behaviors on the part of viewers, in dealing with the initial channel/program selection they make. This study dealt with three of them—knowledge about what was available to watch before the set was turned on, the extent to which and the manner in which television channels were checked at the onset of viewing, and the use of printed guides to assist in the viewing selection process.

Playboy subscribers know what they will watch on television in general before turning the set on significantly more often than they know what they will watch on the PC. One-third of the respondents indicated that, at least 3/4 of the time, they knew what they would watch on television before they started watching, as compared with one-fourth who indicated that they knew in advance what they would watch on the PC that often. Thus, watching on PC was more haphazard or at least fortuitous than the general program-selection pattern of the same respondents.

Checking different channels to see what was on also occurred more frequently with regard to television in general than to the PC. For general television, 40% indicated checking different channels all the time or 3/4 of the time, as compared with the one-third who indicated that they checked the PC to see what was on that regularly.

In addition, respondents were queried as to the specific pattern they used in checking channels when they turned the TV on. From some prior research, four distinct patterns had emerged, and these were examined with this sample. Half the time or more often, these outcomes were obtained:

- 47% checked channels in *numerical order*, beginning with 2 or 3 and then going to 4, 5, 6, etc.
- 40% checked channels in some *idiosyncratic order*, other than 2, 3, 4, etc.
- 57% changed channels (regardless of the order they used) and *stopped changing* at the first "good" program they found.
- 53% checked a substantial number of channels and then *went back* to the best one they had identified.

The first two of these patterns co-occur with the third and fourth. The typologies from these patterns complement the viewers who know in advance what they will watch (61% indicate that is their mode at least one-half of the time). Clearly, the viewer is not constrained to any one of these modes, and no pattern dominates most viewers most of the time.

Use of a television guide to plan viewing was examined—for immediate viewing, same-day viewing, and subsequent viewing. Use of a guide when first sitting down to watch television was significantly more prevalent for general television viewing than for specific PC viewing. Nearly half the sample said they checked a guide when they first started watching television at least 3/4 of the time, in comparison with one-third who said they checked the Playboy section of the guide that often. Thus, paying for a premium channel (or at least, this premium channel) does not lead to as much guide use as does watching television in general.

Same-evening viewing was also more likely planned with a television guide for non-PC programs than for PC ones. Using the guide for planning the evening's general viewing was done "very often" by 30% of the sample, as compared with 21% who used a guide to watch something on the PC the same evening.

Just about one-fourth of the sample very often used the television guide to choose something for viewing later that week or month, in terms of both PC and non-PC programs. So a television guide is used as more of a today-only utility; after all, it will still be available the next time it is needed. Most guides (save those distributed by cable firms) do not go beyond the present week before being replaced, so longer-range planning has not been provided for in the system.

In sum, PC subscribers approached that channel differently than TV in general. They plan their viewing on that channel less; they are less likely to know what they will watch before they turn the channel on; they are less likely to use a guide when they first start watching, or to select something to watch on that channel the same evening. And they are less likely to check different channels when they first turn on that channel.

Re-evaluation.   After a program has been selected for viewing, the viewer has a continuous opportunity to retract the selection—by shutting off the TV or by checking other channels to see what else is on, or even by referring back to a guide for additional selection opportunities. Here, we will examine which options are exercised and when—during a program or at commercials.

The results continue in Table 1. Basically, they show no significant difference in the use of these options when watching television in general or the PC specifically. Nevertheless, one can identify the absolute rate at which these re-evaluation patterns emerge.

Beginning with re-evaluation behaviors *during commercials*, neither guide checking nor channel switching are majority behaviors that are practiced with any considerable regularity. Yet they are reported to occur often enough to strike some fear into any media time buyer for an advertising agency, or salesperson at a commercial network. The average level reported for guide checking during commercials is that it is done about one-fourth of the time; 8% do it almost all the time and 45% do it almost none of the time. Alternate channel checking is somewhat more often exercised during commercials, although the average level of doing it is again one-fourth of the time. At the high extreme, 8% do it almost all of the time, but there is a larger middling group, because only 33% do it almost

Table 1.

		TV	Playboy
1.	Orienting Search, Guide Use, and Planning		
	Know what to watch before set is on	3.0	3.4*
	Check channels when first watching	2.9	3.2*
	Use guide when first start watching	2.7	3.2*
	Use guide to select the same evening	2.2	2.5*
	Use guide to select later in week	2.4	2.4
2.	Re-evaluation		
	Zapping		
	Guide checking at commercials	3.9	4.0
	Channel switching at commercials	3.7	3.8
	During Shows		
	Guide checking during shows	4.3	4.3
	Channel checking during shows	4.1	4.0
	Multiple show watching	3.9	3.9
	Program Completeness		
	Watching shows from start to finish	2.4	3.0*
	Begin watching after show starts	3.4	3.3
	Stop watching before ending	3.7	3.6

*$p < .05$ between the column pairs.

none of the time. So there is a sizeable minority who regularly engage in zapping commercials and checking guides specifically during commercials. The same proportions check the Playboy channel or check the Playboy section of the television guide to see what's on during commercials, as check other channels or other entries in the guide.

Re-evaluation during shows occurs less frequently than at commercials. Asked how often they checked a television guide in the middle of a show other than at commercials, only one-sixth of the respondents did so half the time or more frequently, and the average response was between one-fourth of the time and almost none of the time. Asked how often they checked different channels during the show, one-fourth said they did so at least half the time, a more frequent rate than for guide checking.

Another mode of re-evaluation during shows rarely studied is that of attempting to watch more than one show concurrently. Although, conceptually, this is not necessarily reevaluation, the phenomenon has been too little studied to sort out discontent as a motive from a genuine interest in multiple programming occurring simultaneously. The remote-control channel changer enables the viewer to switch rapidly among channels, so that two programs can be followed, or perhaps more, depending on the program content involved. With cable, virtually all systems have made remote-control devices standard equipment, either wired or wireless. Here multiple-show viewing was attempted by 30% of the viewers at least half of the time that they watched television, and a nearly identical proportion watched something on the PC and some other channel at the same time with that regularity.

In sum, once the PC has been chosen to watch, subscribers are just as likely to zap commercials or programs, either by checking channels or looking at a guide, as they would watching other channels. A PC show is as likely a candidate for multiple-show viewing as programs on other channels.

The final category of re-evaluation behaviors dealt with has been tentatively labeled "program completeness"—whether shows are watched from start to finish, or started after they have begun, or stopped before they are finished.

Television shows are more likely to be watched from start to finish than in any other pattern. However, this sample reported that they were significantly less likely to watch PC offerings from beginning to end than they were to watch other shows on television. Fifty-five per cent typically watched other television shows all the way through at least three-fourths of the time; only 36% watched Playboy in that manner. Perhaps because offerings on PC are repeated, the viewer is not so inclined to view them all in one sitting.

What about other starting and finishing times? Second most common was beginning a show after it had started. Twelve per cent indicated that was something they did almost all the time, and an additional 8% said it occurred 3/4 of the time for them. Least common was not finishing watching a show, 7% did this almost all the time, and 10% indicated they did it 3/4 of the time. For neither of these latter two behaviors did it matter if they were watching television in general or PC movies or other PC programs.

Repeat viewing.   A related viewing phenomenon is that of repeat viewing, watching exactly the same show more than once. Unfortunately, this item was not included for general television viewing, only for viewing of PC offerings. Respondents were asked separately about movies and other programs on the PC, but the results are the same. At least for the offerings on this premium channel, 6% of the subscribers watched the same offering "many times," 17% "several times," 42% "a few times," and 36% said they don't do it.

Channel repertoire.   One outcome of this viewing process has been the determination of the size and composition of the channels regularly viewed by cable subscribers. Typically, they have been given a list of available channels on their system and asked which they watch regularly. This aided-recall method typically yields 10–12 channels selected in 36-channel systems. In this study, an alternative method was used—asking them to identify the channels they watched at least once a week, with no list to prompt them. Given that testing was done in four different systems, with different channels available, interviewers had a master list of 42 cable network channels which they checked off as specific channels or networks were cited. On average, the respondents spontaneously identified 4.0 cable channels: 14% named one, 13% named two, 20% named three, 15% named four, and no other frequency exceeded 10% of the sample. Information about noncable channels cited, primarily commercial network stations, was not systematically recorded. Nevertheless, it is clear that unaided recall provides a smaller set of channels in the channel repertoire of the individual viewer, and that the number identified is a small fraction of the full set available to view.

Viewing models.   Two simple models of program choice were created from the data already discussed, one for general television viewing and one with PC viewing behaviors. It is instructive to compare these two models in Figure 1. The *planning* variable listed combines guide-use items and knowing ahead of time what will be watched. *Orienting search* refers to the tendency to search channels before deciding what to watch. *Re-evaluation* is an index of zapping items at commercials and during a show. *Channel repertoire* is the number of cable channels watched at least weekly. Parallel items were combined for testing within the Playboy viewing model.

Within the general viewing model, the orienting search behavior is strongly related to commercial zapping—or channel checking before deciding what to watch goes along with channel switching after watching has begun. None of the other relationships are substantial.

Within the viewing model for this specific channel, the viewing process elements are more tightly packaged with each other, as might be the case whenever we move from the general case to the specific. Planning is accompanied by more orienting search activity, both of which precede more re-evaluation. The active planner is an active chooser and an active re-evaluator of programs.

Figure 1.   Viewing Models

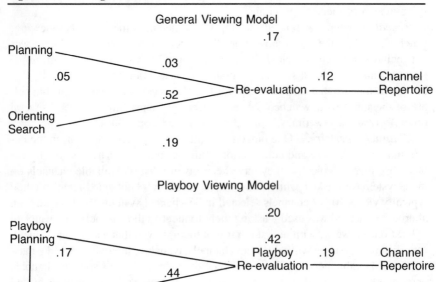

Satisfaction and Use of the Playboy Channel

The final analysis to be reported is channel specific, presented as a prototype analysis for other premium channels. The research question was that of predicting use of the channel and satisfaction with it, based on viewing-style variables, in addition to the more typical set of demographic attributes of premium-channel subscribers.

Use was assessed by asking subscribers if they had watched each of 25 different PC programs (by title) during the previous 30 days. The average number viewed was 6.7, or more than 25% of all the listings. There was high variability in the usage distribution:

Ten percent watched none of the shows, 2% watched 1, 9% watched 2, 8% watched 3, 6% watched 4, 8% watched 5, 11% watched 6, 10% watched 7, 8% watched 8, and 28% watched >8.

For tapping satisfaction with the PC, four items were used. These three had four response categories ranging from Agree Strongly to Disagree Strongly:

- Overall, I am satisfied with the Playboy Channel. (74% agreed)
- I plan to continue my subscription to the PC. (81% agreed)
- The PC gives me my money's worth. (62% agreed)

The fourth item was:

- Overall, how would you rate the PC? (On a five-point scale, 45% were on the positive side of liking, 38% were at the midpoint, and 17% were on the negative side.)

These four items were summed in a composite measure of satisfaction. The average score was 12.2, with 17 the maximum positive response.

Three sets of variables were intercorrelated with the constructed indices of use and satisfaction. The three sets were:

1. Demographic characteristics: Age, marital status, number of children, income, gender, and number of pay channels (other than PC).
2. Viewing style variables: Here, several constructed viewing-style components were used. These included estimates of *re-evaluation* (changing channels during commercials and during a show other than at commercials), *checking channels* when the set is first turned on, *preplanning* (general guide use and planning ahead to view), and *searching* (the extent to which the viewer searched channels before deciding what to watch). Recall that estimates of re-evaluation and preplanning were asked twice—once for general television viewing and once for PC viewing. A duplicate analysis was done using those channel-specific estimates, plus the channel-specific measures of watching PC shows from start to finish and stopping one's viewing before the show ended. The general viewing analysis will be discussed in detail; comparative statements will be made regarding the second analysis.
3. Use, satisfaction and channel repertoire (number of different cable channels viewed at least weekly) were also included.

Table 2 presents the simple and multiple correlations of these three sets of variables with use of, and satisfaction, with this channel, for both the general and specific situations.

Satisfaction.   First, the general viewing-style variable results. Correlations were found for age and sex with channel satisfaction. In this instance, it was younger and male subscribers who were more satisfied. As for viewing style correlates of satisfaction, those who did more active viewing—more re-evaluation during commercials and programs, more channel checking when they made their initial viewing decision, and searching in some specific search pattern—were more satisfied consumers. Finally, watching more different programs on the channel, and channel repertoire, were significantly correlated with satisfaction.

The multiple regression analysis provides additional parsimony to the findings. The group of demographic variables had a multiple correlation of .37, adding the group of viewing-style variables increased that multiple correlation to .41, and adding use and channel repertoire yielded a final R = .48. In that analy-

Table 2. Correlates of Satisfaction and Use of Playboy Channel

	General Viewing		Playboy Viewing[b]	
	Sat.	Use	Sat.	Use
Demographics				
Age	−.13	−.05	−.10	−.11
Not single	−.00	−.01	.01	.07
No. of children	−.01	.03	.06	.10
Education	−.02	−.01	−.03	.03
Income	−.07	−.07	.02	−.09
Gender	.32*	.28*	.34*	.31*
No. of pay channels	−.01	−.06	.08	−.02
	(.37)[a]	(.31)	(.37)	(.37)
Viewing Styles				
Reevaluation	.18	.22	.38*	.34
Channel-checking	.17	.25	.22	.28
Search Pattern	.24*	.22	.28*	.28
Preplanning	−.04	.03	.25	.42*
Playboy Stop Watching	—	—	−.07	.00
Playboy Finish Watching	—	—	.32*	.41*
	(.41)	(.39)	(.56)	(.60)
Outcomes				
Satisfaction	—	.34	—	.39*
Use	.34	—	.39*	—
Channel Repertoire	.17	.26*	.17	.12
	(.48)	(.51)	(.58)	(.60)

[a]Multiple correlation (cumulated) for each group of variables
[b]These conditions vary slightly from Table 1 because this analysis was conducted on the 148 respondents for whom Playboy information was available in a guide.
*Significant Betas

sis, the specific variables which were the primary significant predictors were age, gender, use, and searching channels in some specific pattern.

Use. The second column in Table 2 provides parallel findings for the measure of channel use. Variables which individually correlated strongly with use of PC were gender (male), frequent reevaluation, more channel checking, having more of a specific search pattern, watching more shows on the PC, and having a larger channel repertoire. Not surprisingly, the list is almost identical to that obtained with satisfaction as the dependent variable. Age, the weakest correlate in the prior set, dropped away as a predictor of use.

The collective strength of these sets of variables is displayed in the multiple regression results. The demographic variables had a multiple R = .31; adding the viewing-style variables increased it to .39, and then adding satisfaction and channel repertoire boosted it to .51. The primary predictors were gender, satisfaction, and channel repertoire. So the estimates of use and satisfaction were predicted to the same extent and by the same subset of variables.

When the channel-specific viewing style variables were substituted for the nonspecific ones, the importance of the viewing-style variables increased substantially, in terms of both individual correlations with use and satisfaction, and their contribution to the multiple correlation results. (See Table 2.) As evident, the demographic characteristics relate in essentially the same way, despite some difference in sample composition in the two analyses; altogether, they each provide an R = .37. Gender accounts for virtually all of the variance attributable to the demographic variables. However, adding the specific viewing-style variables increases the multiple R for satisfaction by .19 (to .56) and for use by .23 (to .60), (as compared to increases of .04 and .08, respectively, when using the general viewing style measures). Here, in this second analysis, all the specific viewing-style variables—re-evaluation, channel checking, searching in some specific order, prior planning, and watching shows until they finish—save one, are strong, significant individual correlates for both use and satisfaction. Several remain significant in each multiple correlation analysis, namely, re-evaluation, search pattern, preplanning, and watching shows until they are finished, in addition to gender. Then, adding the outcome variables does not add substantial predictive power. Thus, the viewing style variables are impressive correlates of satisfaction and use in both the general and specific case, but substantially more so in the latter.

DISCUSSION

Differences do emerge between what are reported as general viewing-style behaviors and those same behaviors in the context of a specific-content channel. It would be anticipated that differences in the same or other behaviors would be found if the programming content were of a different genre than that on the Playboy Channel, e.g., watching news on CNN, or a congressional hearing on C-SPAN. And it would be anticipated that watching similar content across channels would vary less, e.g., watching sports on ESPN and sports on CBS, or movies on HBO and movies on the Playboy Channel.

What has been demonstrated, rather than speculated about, is that asking about viewing of the content on a specialty-content channel results in greater reliability among these viewing style factors and variables. The two simple configurations of intervariable correlations showed that several of the behaviors were more closely associated in the more specific-content viewing situation. More importantly and impressively, indices of the viewing measures were significant correlates of certain outcome behaviors of substantial interest—channel usage and channel satisfaction—*only* in the situation where those viewing measures reflected specific viewing approaches to the channel. Where they were more general estimates of behavior, the strengths of the correlations were, on average, half as great. For the variables used in common between the specific and general viewing situation, the lowest of the Playboy viewing-style variable correlations was equivalent to the highest of the general viewing ones.

So, we begin to develop the notion that the viewing-style variables are sensitive to content differences, that the way one form of television is watched is not necessarily equivalent to the way in which another is followed, that there is reciprocity between the viewer's propensities and the perhaps inherent demands of the content on that viewer. The origins of the behaviors remain an unknown. Are they merely our means of coping with television, our means of adapting to it? Have we some sense of greater control by us over the medium? Have we modeled peers or parents in developing these patterns?

Before questioning too far, this study and the data in general on viewing style variables would have been enhanced greatly had an experimental research opportunity been equally available. Then, groups of individuals could have been exposed to Playboy Channel offerings. In that context, they also would have had available a full array of alternative cable system channels, but perhaps instructed to "try to follow what's on the Playboy Channel." They would be observed during these sessions, and to the extent the observations corresponded to their reports of their own behavior, some verification of the self-report procedure would be available. But this experimental process is tainted, because the viewing environment is not one's home and knowledge about being observed may alter behavior more considerably than content considerations. An alternative, now being used by the Independent Broadcast Authority's research department in England, is the placement of an unobtrusive camera in the home (enclosed within a piece of furniture, on top of which the TV receiver is placed) monitoring viewing patterns. After one or two sessions, inspection of the resulting tapes indicates very little if any sensitivity to the camera's presence. Although there may be some greater sensitivity to viewing of the Playboy Channel, in particular, it is unlikely that viewing CNN, HBO, MTV, etc. would be unresponsive to the data collection procedure.

And of course, the special content of the channel studied makes generalizations to other specialty-content channels weak. Replication with more widespread specialty channels is needed to confirm the phenomena first studied here.

Still, the magnitude of the behaviors reported is curious. There is much activity preparatory to watching television—channel-checking, guide use for the same day, and even knowing what to watch before the set is on, all occur fairly regularly. By contrast with these orienting-search behaviors, there is much less commercial zapping, or even channel switching at commercials, or multiple show watching. Most typically, these adults report watching shows on this channel from start to finish, rather than starting or stopping precipitously or moving around channels while the show is on. This is not to say that this more active viewing mode for re-evaluation is not exercised; it is just that it goes on "about one-fourth of the time" rather than more often than that.

Our speculative explanation for this is that these alternative viewing styles—more checking, changing, dual watching, guide use—are learned behaviors, and these viewers learned them somewhat later in life. Cable was added to their media collection, as were switching devices, video recorders, etc., much after they "learned" to watch television. Although it is now relatively effortless for them to alter their television viewing process, they may be slow to do so. At any rate, if this explanation has any credence, the proof of it will be in the next generation of viewers, the offspring of this sample. For them, the image invoked by the concept of television has *always* encompassed 36 channels, a remote-control device of some sort, a companion VCR for taping a second program, the capability of screening 36 channels in less than 30 seconds, etc. Few of these are novelties for viewers under 15 years of age, and, in another half-generation, will likely be novelties for a minority of the U.S. population. Thus, one might expect to find a generation of more active viewers who have developed a normative viewing style that greatly increases the magnitude of the behaviors reported here. They may zap more, change channels and programs more readily, more often watch multiple shows concurrently, etc. Whether this viewing style is then characterized as active, or frenetic, or haphazard will depend on the outcomes yet to be determined from watching television in this manner. Will they learn less, enjoy it less, recall it less, watch fewer options—or will their increased experience with and capacity for absorbing television material enhance these same responses? This could well be studied currently with much younger viewers than in this project; that might be a meaningful way of anticipating the form and pattern of tomorrow's viewing styles.

SECTION FIVE

Subscriber Types

CHAPTER 16

Cable and Noncable Viewing Style Comparisons

Bradley S. Greenberg, Carrie Heeter,
David D'Alessio, and Sherri Sipes

Cable subscriber status has been a key discriminator used in much of the cable audience research available to date. Some research compares cable with noncable households. Other studies distinguish pay and basic households. The nearly exhaustive list of subscriber types which might be contrasted would include: *broadcast households* (homes which have not been passed by cable, and therefore cannot subscribe), *nonsubscribers* (who have had the choice and decided not to subscribe), *basic subscribers* (who pay only the minimum cable rate, and receive no pay channels), *single-pay subscribers* (who pay for one extra cable channel), and *multipay subscribers* (who pay for two or more special cable channels above and beyond basic cable). Then there are all the "ex" subscriber categories, those who gave up all of cable, either voluntarily or through disconnection, or those who gave up one or more pay channels, or changed pay channels, etc. Of these groups, the only one not growing is broadcast households. At least 9% of nonsubscribers once subscribed to cable (LeRoy & LeRoy, 1983), and that proportion grows each year. Among nonsubscribers, 22%–49% say they will eventually subscribe (Reymer & Gersin, 1982). Twenty percent of basic subscribers have cancelled a pay service, and more than 30% of pay cable homes subscribe to more than one pay channel (LeRoy & LeRoy, 1983).

Are there reasons to group individuals by subscriber status? Is there a foundation for presuming similarity between basic cable subscribers in a small cable system who receive seven channels, two of which are not available over the air, and basic cable subscribers in a 54-channel system who receive 40 channels, 33 of which are not available over the air? Or is there merit in proposing that these basic subscribers are fundamentally different in (a) their television orientation, (b) their media environment, or (c) the impact television has on them as compared to their pay cable counterparts who pay an additional fee to receive one or more additional channels.

This chapter will critically review research distinguishing different subscriber groups and present original data that focuses on two areas: background attributes, primarily demographic characteristics which relate to the decision to subscribe; and viewing behaviors (which may be either outcomes or causes of the decision to subscribe). The importance of subscriber status in predicting viewing behaviors is then considered.

Several factors should act as cautions for subscriber status studies in general, all with a central theme—cable systems vary. The systems themselves differ in terms of how many channels they offer for basic service, how many pay channels are available, how many tiers of service are offered, the rates for basic pay and tiers of channels, and whether or not remote-control channel selectors are available and at what cost. In addition, the existing broadcast media environment varies: large markets receive as many as 10 or more channels off the air without cable television. The smallest markets may receive only one or two broadcast channels, without cable and often with poor reception.

Further, the age of the cable system makes a difference in at least two ways. First, subscriber penetration rates tend to increase over time, up to a saturation point, such that the subscriber profiles would be different early and late in a cable system's life cycle (Sparkes, 1983a). Churn (disconnects) continues at high levels (as much as 30% of cable subscribers stop subscribing per year), further altering subscriber–nonsubscriber group composition. Second, cable television itself has changed dramatically in its brief history. Pre-1980 studies were conducted before the establishment of cable satellite networks and therefore were describing a very different television service. Here, only studies conducted after 1978 will be reviewed.

Demographics

Demographics have been examined frequently for differences among subscriber groups. Webster (1983), collapsing across 14 markets, found a significant nonlinear age relationship, with basic subscribers the oldest group, pay subscribers the youngest, and nonsubscribers in between but closer to basic ones. Two studies, also across multiple markets, found pay to be younger than basic (Webster & Agostino, 1982; Ducey, Krugman, & Eckrich, 1983). Sparkes (1983b), in analyzing nonsubscribers across three different studies, suggests that there are two distinct age groups of nonsubscribers: those of retirement age, and those 20 to 30 years old. TAA (Television, 1983) also found that more 65 + respondents were nonsubscribers, while Rothe, Harvey, and Michaels (1983) found cable subscribers to be younger. In general, we expect:

H1:   Pay cable subscribers are younger than basic or nonsubscribers.

Income is generally found to be higher among cable subscribers (Henke, Donohue, Cook, & Cheung, 1983, LeRoy & LeRoy, 1983; Sparkes, 1983b;

Grotta & Newson, 1981, Rothe et al., 1983). In addition, pay subscribers have higher incomes than basic subscribers (Ducey et al., 1983). Sparkes (1983a) suggest that income varies with the age of the cable system, demonstrating that low income residents tend to subscribe early but also tend to disconnect early.

> H2:   Pay cable subscribers will have higher income than basic cable subscribers, who will have higher income than nonsubscribers, at least in mature systems which have been operating for several years.

The presence of children and size of household appear to be related to cable subscription. Cable households have larger families (Television, 1983; Webster & Agostino, 1982; Sparkes, 1983, Rothe et al., 1983). Pay households are larger and more likely to include children than either basic (Ducey et al., 1983; LeRoy & LeRoy, 1983) or noncable (Webster, 1983). However, Webster found a nonlinear relationship for household size, such that pay had the largest, basic the smallest, and broadcast an in-between household size. Sparkes (1983), in a survey of nonsubscribers, found that 90% lived in homes with two or fewer adults and no children. Among households with children, Sparkes reported a 77% penetration rate for cable in one system, compared to 50% in households without children. Across the different studies, it is not clear whether respondents were asked about number of children or the number living at home. So, we expect:

> H3:   Pay cable households will have more children at home and larger households than basic or noncable households, with noncable being the smallest.

Findings for education as an attribute have been the most contradictory. Education was found not to be a significant discriminator between cable and noncable, or between basic and pay in both four-market (Ducey et al., 1983) and single-market (Collins, Reagan & Abel, 1983) studies. Sparkes (1983) found that higher education was the strongest predictor of subscribership to cable early in the life of a cable system, but dropped below income and household size later in the system's history. Again, nonsubscribers appear to be composed of two subgroups. TAA (Television, 1983) reports that nonsubscribers constitute more college graduates and more high school dropouts than cable subscribers, in two markets. A single market study by Henke et al. (1983) reports that cable subscribers had higher education (42% of subscribers and 17% of nonsubscribers were college grads). LeRoy and LeRoy (1984) conclude that pay cable subscribers have more education. Given these inconsistencies, we do not derive a hypothesis for the role of education in cable subscription.

One basis for presuming a difference related to subscribership is the possibility that subscribers are more innovative. Krugman and Eckrich (1982) characterize pay subscribers as more impulsive, more likely to be opinion leaders, innovative, and new brand triers, compared to basic subscribers who are more deliberate and traditional. Krugman, Eckrich, and Ducey (1983) found that pay subscribers also were more likely to own TV games and VCRs; TAA (Television, 1983) also reports greater VCR, videogame, and TV ownership among pay subscribers than among other categories, and Rothe et al. (1983) report more VCRs, videogames, and home computers in cabled homes, all of which is likely an income artifact. Therefore, we expect:

> H4:   Pay subscribers will own more electronic/TV gadgetry than basic or non-subscribers, and will have more innovative attitudes.

Viewing Behaviors

Turning now to viewing behaviors, there likely is some difference in the amount of television viewed by subscriber status. Nielsen reports that pay households view an average of 20% more hours per week than nonsubscribers (in LeRoy & LeRoy, 1983). Webster and Agostino (1982) report that pay watch more than basic, in four markets, using Arbitron diary data. On the other hand, in four other markets, using mail survey data, Ducey, Krugman, and Eckrich (1983) report no significant difference. Webster (1983), across 14 markets, finds that pay cable homes watch more hours than basic, who watch slightly more than nonsubscribers. But the fact that pay households are larger may explain some of this difference. TAA (Television, 1983) reports a slight difference for pay, particularly multipay households, as compared to nonsubscribers. Collins, Reagan, and Abel (1983), in a single market in a new system, found lower viewing among cable than noncable households. Sparkes (1983b), in a single market, found equivalent cable–noncable viewing levels in his first survey, and an increase in cable viewing 18 months later. The weight of the evidence, less than definitive, suggests that:

> H5:   Pay cable subscribers watch more television than basic or nonsubscribers.

A sidebar on this hypothesis is worth specifying. Webster (1983) demonstrated that viewing shares devoted to different channel types varied according to the broadcast market and the availability of off-air signals. The general trend is for broadcast households to watch the largest proportion of local station programming. Basic cable subscribers tend to watch more distant stations (also available off-air) than pay subscribers, who instead are watching the pay channels. Other cable-only channels are watched at very low levels, as a whole comprising between 1% and 7% of all viewing, in both basic and pay households. From Webster, we derive:

> H5a: Cable subscribers will watch more distant off-air channels, with basic exceeding pay subscribers, whereas pay and basic cable subscriber consumption of cable-only channels—exclusive of pay channels—will not differ markedly.

One clear difference in viewing behavior by subscriber status is the number of different channels watched. Cable subscribers watch more channels than nonsubscribers. In two markets; nonsubscribers watched an average of four different channels in 2 weeks compared to six channels in one of the markets and 7-plus in the other for cable subscribers. (Television, TAA, 1983). Nine of 32 basic cable channels and eight of 28 were watched by one-third or more of cable subscribers on at least one evening in 2 weeks. The actual level varies by cable system. In systems with seven to eight cable channels, an average of four different channels are watched; with 11 to 15 available, six are watched; and with 26 available, 10 were watched (LeRoy & LeRoy, 1983). Webster (1983) and Webster and Agostino (1982) both report that pay subscribers watch more different channels than basic. Webster (1983) reports the average across 14 markets; nonsubscribers watch an average of 3.95, basic subscribers 6.05, and pay subscribers 7.68.

> H6:   Pay subscribers watch more different channels than basic subscribers who watch more different channels than nonsubscribers.

On the average, viewers know what they will watch about half the time, and, the other half of the time, they decide what to watch at the time of viewing (Television, 1983, chap. 2). TAA found cable viewers neither more nor less likely to plan their viewing in advance. They conclude that cable seems to affect how, but not when, viewers decide what to watch.

> H7:   Planned viewing will not differ by subscriber status.

Findings about guide use are contradictory. Sparkes (1983b) reports no significant difference between cable and noncable use of guides—although 88% of the cabled respondents report very often or occasionally using the channel which shows a program guide. But how typical are systems with on-air cable guides? TAA (Television, 1983) reports small differences among groups: 52% of the nonsubscribers were regular guide users, compared with 55% of those with basic cable and 62% with pay channels.

> H8:   Pay cable subscribers will use guides more often than basic or nonsubscribers.

This book develops a model of program choice which differentiates two types of channel changing. One is the *orienting search*, which refers to channel sampling, or the pattern of flipping through program options before a program is selected for viewing. The second is the pattern of *reevaluation* which refers to channel changing either during commercials (typically referred to as *zapping*), or during a program being watched, after the initial decision to watch a show has been made. Reevaluation is related to viewers' commitment to a specific program and to programs in general, and to their propensity to check alternatives after an initial choice has been made. TAA (Television, 1983) reports that 33%

of noncable viewers, 47% of basic-only, 49% of single-pay and 52% of multipay viewers often or always scan channels before deciding what to watch. Nine percent of nonsubscribers and 24% of subscribers had remote control. Presence of remote control made no difference for nonsubscribers in use of orienting searches, but cable orienting searches increased from 47% without remote to 54% with remote. Chapter 2 reports pay cable subscribership to be positively associated with orienting searches (compared to basic), but a weak factor in predicting orienting searches, after controlling for demography and other viewing behaviors.

> H9: Pay cable subscribers will engage in more orienting searches than basic, who will engage in more than nonsubscribers.

For reevaluation, TAA reports that 6% of nonsubs and 17% of subs change channels often or always during a program. Thirteen per cent of nonsubs and 39% of subs do so during commercials. A Neilsen study of dial switching found little difference between noncable and basic household dial switching (4.7% and 4.8% share during commercial minutes, and 2.8 and 3.2% share during noncommercial minutes, respectively). Pay cable evidenced more switching—6.2% share during commercial and 4.0% share during noncommercial minutes. However, this difference may be confounded by the presence or absence of a remote control selector. In the Neilsen study, 14% of noncable, 16% of basic, and 28% of pay viewers had remote control. The dial-switching share for remote control, regardless of subscriber status, was 7.1% during commercial and 4.0 during noncommercial minutes, while, for cable with remote, it was 7.0 and 4.8.

> H10: Pay cable subscribers will change channels more often than basic or nonsubs (but presence or absence of remote control should be controlled).

The listed hypotheses are fairly straightforward comparisons among two or three subscriber status groups, largely bivariate hypotheses. Two additional research questions and their associated analyses will examine interrelationships; these are designed to contribute to a more parsimonious understanding and conception of the cableviewing-subscriber process. First, which among the variables examined here best discriminate respondents by subscriber status, within a multivariate analysis format? Second, how important is subscriber status in predicting these viewing behaviors? Are the differences better explained by factors other than subscriber status, or does it remain a significant predictor controlling for other influences?

METHODS

Households in a midwestern city served by a cable system franchised to a major multiple-system operator were surveyed by phone. The 35-channel system carried two affiliates from each commercial network, two public television stations (one local and one nonlocal), five text channels (e.g., sports, weather), two pay channels (HBO and Cinemax), five satellite cable channels (e.g., ESPN and MTV), 11 channels dedicated to education, government, and public access, one superstation, one independent, one foreign station, and one channel of inspirational programming. Six commercial stations (two affiliates per network), one PBS station, and two UHF independents were available over the air without cable. The cable system had been in operation since 1972, and can thus be considered a "mature" system. Remote control channel selectors were available for an additional $1.50 per month, and about half of basic and pay subscribers elected to have remote control.

Sample.   The phone sample was created by randomly selecting phone numbers with prefixes within the cable system area, using the first six digits of the numbers, and assigning 0 to 9 for the seventh digit. The completion rate was 81%, after eliminating numbers not in use and businesses.

In all, 346 interviews averaging 15 to 20 minutes were completed by nine interviewers in the fall. After training and pretesting, interviewers made calls between 6 p.m. and 10 p.m. on weekdays and on weekend afternoons and evenings. Interviewers first asked for male respondents and then females, 18 years of age or older. The completed sample was 49% male.

Variables.   For the purposes of this project, three groups of variables were examined—background variables, viewing choice process variables, and other viewing behaviors.

Background.   Prior research identified selected demographic characteristics as likely to differentiate cabled from uncabled respondents, and pay cable from basic cable respondents. Included were number of children living at home, household income, educational level, and age. In addition, two measures of "innovativeness" were used: (a) an additive index of ownership of technology items (video games, home computers, video recorders, and videodisc players), and (b) innovative orientation, for which responses were obtained on a five-point agree–disagree scale to eight statements, e.g., "I enjoy trying out new ideas," and "I am suspicious of new inventions." All items in this scale intercorrelated at $p<.002$ and were summed.

Viewing behaviors.   For overall television viewing, respondents were asked, "On a typical week day (morning, afternoon, evening), how many hours of TV do you watch?" Responses to the three dayparts were analyzed separately and as a composite measure. Respondents were further questioned about "daily, weekly, monthly, or less" use of each channel available to them, as identified earlier. Responses were weighted to transform an essentially ordinal response set into one which approximated equi-intervality on a monthly base: daily became 30, weekly became 5, monthly 1, and less was zero. Results are examined in terms of off-air channel use for all three subscriber groups and of cable-only channels for pay and basic subscribers.

Viewing choice process.   This study asked about choice process behaviors in two contexts—viewing alone, and viewing with others. The two sets of measures are correlated and for analysis purposes here, the two contexts are combined where possible.

Choice process behaviors consisted primarily of questions to which the respondents indicated how often they engaged in those behaviors; the scale responses were "almost all of the time, 3/4 of the time, 1/2 the time, 1/4 the time, and almost none of the time." *Planned viewing* was derived from asking how often the respondents knew what they would watch before turning the set on, asked separately for viewing while alone and in a group situation. *Guide use* consisted of the frequency of referring to a television listing before the set is turned on, and while watching (both while alone and in a group). *Orienting search* behaviors were assessed by asking how often the respondent made initial selections by changing channels and stopping at the first show that looked good, watching whatever happened to be on, and changing channels to see what else is on, before deciding what to watch. *Reevaluation* was assessed separately for self- and group-viewing situations. Individual reevaluation came from asking how often the respondent changed channels while commercials are on, between and during shows ($r = .70$). The group context was assessed by asking the same pair of questions, plus the frequency of changing during shows other than at commercials, and the frequency of others changing channels. All items in this index were intercorrelated at $p < .001$). *Channel repertoire* was the summation of all available channels a respondent reported watching on a daily or weekly basis.

Analysis.   Three sets of analyses bear on the hypotheses and research questions raised. First, one-way analyses of variance examine viewing and choice behaviors and demographic attributes by subscriber group. Because many of the hypothesized differences distinguish one subscriber group from the other two, two-group comparisons are also presented where appropriate. Second, discriminant analyses identify variables which maximally differentiate broadcast from cable respondents, and basic from pay cable respondents. Third, in multiple regression analyses, subscriber status is examined, together with background and demographic attributes, to assess their role in predicting TV-viewing and choice-process behaviors. Four variables were used as control variables in the multiple regressions: presence or absence of remote control, proportion of viewing time spent watching alone and with children, and the frequency with which the respondent reported participating in program selection during group-viewing sessions.

RESULTS

Group differences.   Table 1 identifies subscriber group differences in background variables. The sample is presented in terms of subscriber group membership—broadcast-only, basic cable-only, and pay-cable status. The statistical results provide first the overall comparison among the three groups and then for pairings of subscribers.

Background variables.   Several hypotheses involving background or demographic comparisons are being replicated largely for confirmatory purposes, and with somewhat greater specificity perhaps than before; we will report on those first.

The youngest subscribers were the pay cable subscribers, averaging 35 years; next were the basic cable subscribers, and finally the broadcast-only group. As hypothesized, the pay-cable subscribers were younger than the nonsubscribers, and younger than the nonpay subscribers, taken altogether; the pay-basic difference was directionally as predicted but not statistically significant in this sample. Further, cable subscribers were much younger, on average, than nonsubscribers. Hypothesis 1 was partly supported.

Income differences were apparent across the subscriber groups. Pay cable subscribers had higher average incomes than basic subscribers and higher incomes than nonsubscribers. This was a consistent linear trend across the three groups. Hypothesis 2 was supported fully.

Size of household was examined in terms of number of children in the home. This factor also differentiated the pay-cable households from the others. On average, the pay-cable subscribers had one child per household and the other two groups averaged one child for every two households. However, the basic subscribers were not different from the nonsubscribers in terms of number of children. Hypothesis 3 was partly supported.

We deferred from predicting educational differences because of the inconsistency of prior research in regard to this characteristic. The present study is supportive of minimal or no educational differences. The trend line indicated that the least-educated group was found among the pay cable subscribers, but that trend was not significant.

Table 1.   Background Variables by Subscriber Status

Variables	B-cast (n = 133)	Basic (n = 81)	Pay (n = 127)	Overall	Cable–No-Cable	Pay–Basic
1. Age	41.2	39.0	34.8	.01	.02	ns
2. Income[a]	1.15	1.31	1.68	.01	.02	.04
3. # of children	.47	.51	.80	.01	ns	.03
4. Education[b]	6.02	5.63	5.50	ns	ns	ns
5. Ownership[c]	.18	.29	.51	.01	.01	.04
6. Innovative Attitudes	4.31	3.58	5.22	.03	ns	.02
7. Remote control	13%	50%	47%	.00	.00	ns

[a]Income categories were: $45,000+, scored as 4; $35–$44,000 = 3; $25–$34,000 = 2; $15,–$24,000 = 1; less than $15,000 = 0.
[b]These scores reflect an average of 13–14 years of education.
[c]Sum of yesses to ownership of videogames, home computers, videodisc players, and videocassette recorders.

Two measures were utilized to tap the characteristics of innovativeness—one behavioral and one attitudinal. The behavior measure consisted of the acquisition of a variety of electronic communication gadgets. The ownership index constructed indicated highest levels of acquisition of these items among the pay-cable subscribers, followed by the basic subscribers and trailed by the nonsubscribers. Each paired comparison was statistically significant. The measure of innovative attitudes revealed significant findings in accord with what was hypothesized. Most receptive to new ideas were the pay-cable subscribers, who were more receptive than either of the other subscriber groups. However, the basic-cable and noncable groups did not differ. Hypothesis 4 was supported.

Viewing behaviors. Table 2 presents the major findings for each of the viewing behavior variables used. Hypotheses 5 and 6 deal with amount of viewing, total channels viewed, and subsets of channels viewed. As for overall viewing, it was predicted that maximum viewing of television would be found among the pay-cable subscribers. The findings support that contention in part. Broadcast respondents watched less television (3.7 hours per day) than cabled respondents (4.4 hours), but the pay-basic subscriber groups watched the same average daily amounts. The original measure of viewing was for three day-parts; although the average viewing time of cabled respondents was higher in each day part, it was significantly higher only in the post-6 p.m. viewing. So the overall difference in total viewing identified really consists of more viewing in the evening in cabled homes. Hypothesis 5 is supported in part.

What channels are watched, and in what quantity? First, in terms of channels equally available in cable and noncabled homes—the off-air channels—there was more viewing of these channels on a typical day by those with cable than by those without. Breaking down these off-air channels into local affiliates, distant affiliates, and the PBS stations, the locus of the difference, as hypothesized, was the distant network affiliates. Whereas cabled homes typically watched two of

Table 2.   Viewing Variables by Subscriber Status

	Status			(p-values)		
Variables	B-cast (n = 133)	Basic (n = 81)	Pay (n = 127)	Overall	Cable–No-Cable	Pay–Basic
1. Time (hours)	3.67	4.35	4.43	.03	.01	ns
2. Off-air[a]	4.2	4.9	5.1	.00	.00	ns
local affiliate	2.6	2.5	2.6	ns	ns	ns
distant aff.	1.0	1.9	2.0	.00	.00	ns
PBS	.5	.5	.5	ns	ns	ns
3. Cable-only	—	4.8	6.2	—	—	.03
text	—	1.0	1.6	—	—	.01
cable nets	—	1.4	1.7	—	—	.09
distant channels		2.0	2.1	—	—	ns

[a]Numbers of channels indicated under off-air and cable only headings reflect the average number of channels of this type which respondents reported watching daily or weekly.

these distant affiliates, the noncabled homes watched one on average. Thus, this constituted additional channel viewing in cabled homes, rather than diversion from local stations, at least in terms of number of channels, and is likely reflected in the overall viewing figures.

Among channels available only to cabled respondents, pay subscribers watched more cable-only channels, excluding pay options; pay subscribers regularly used 6.2 cable-only channels, compared to 4.8 for basic subscribers. Pay viewers were particularly more likely to use the text channels and cable networks, with no difference in use of alternative public television stations.

Thus, there was support for the notion (Hypothesis 5a) that cable subscribers would watch more distant channels, but not that basic would do so more than pay subscribers. There also was support for the notion that pay subscribers would use more different cable channels (not counting the pay channels subscribed to) than the basic subscribers, and that both would regularly watch more different channels than uncabled households (Hypothesis 6). However, an even stronger indication of the diversity of channel use in cable households comes from the measure of total different channels watched from among all those available to the different subscriber status groups. This measure, of *channel repertoire*, is listed among the choice behaviors in Table 3. Pay viewers reported watching 12.5 channels, basic reported 9.8, and broadcast-only 4.5. Note that the pay viewers regularly watched nearly three more channels on a daily or weekly basis than basic subscribers, in a system with only two pay channels available.

Choice process behaviors.   Next, we turn to the viewing process variables specified. Planned viewing—knowing beforehand what will be watched—had received contradictory outcomes in research reviewed, and our own hypothesis (H7) was one of no difference. And indeed, there were no differences in planning what to watch among the three groups in this study, nor between any of the pairings.

Guide use, on the other hand, did differ by subscriber status. Specific guides generally have been part of the package offered to pay subscribers, and provided

Table 3.   Viewing Process Variables by Subscriber Status

	Status			(p-values)		
Variables	B-cast (n = 133)	Basic (n = 81)	Pay (n = 127)	Overall	Cable–No-Cable	Pay–Basic
1. Planned viewing[a]	2.78	2.60	2.62	ns	ns	ns
2. Guide use	1.32	1.53	1.83	.00	.00	.05
3. Orienting searches	1.09	1.23	1.41	.03	.04	ns
4. Alone reevaluation	.67	.98	.91	ns	.03	ns
5. Group reevaluation	.81	.93	.91	ns	ns	ns
6. Channel repertoire	4.45	9.84	12.45	.00	.00	.00

[a]Respondents were asked how often they engaged in each of the choice behaviors, with responses scored as: Almost all of the time = 4; ¾ of the time = 3; ½ of the time = 2; ¼ of the time = 1; almost none of the time = 0.

to basic subscribers as a marketing tool to introduce them to what is available on the pay channels they have not yet chosen to subscribe to. Guide use was found to be at the highest level among pay subscribers, next among basic, and least among noncabled respondents. The overall difference and the paired comparisons were all different statistically, in support of Hypothesis 8.

Orienting searches, or the extent to which the viewer sorted through more of the available channels before deciding what to watch rather than watching whatever happened to be on, were significantly more common among the cabled respondents than uncabled, and primarily among those with pay channels. Pay and basic viewers, however, did not differ strongly in the magnitude of this behavior. The findings support much of Hypothesis 9.

Reevaluation—or "zapping," as it has come to be more popularly known—was a more common occurrence in cabled households than in uncabled, but only in solo viewing situations; the group viewing means follow the same trend, but not substantially enough to be significant. Interestingly, the absolute levels of change in the group viewing situations are as large as in the solo viewing situations, and appear to be larger among broadcast homes in the group viewing situation. Hypothesis 10 then is supported for the situation in which television is watched while alone. As for the importance of remote control selectors, half of the cabled homes reported having them, compared with 13% of the noncabled homes.

Discriminant comparisons. Two discriminant analyses will be reported, the first comparing cable with noncable respondents, and the second comparing basic with pay cable subscribers. Inputted were the background variables, estimated television viewing time, and the viewing process behaviors. Because number of channels and access to a remote-control channel selector were differentially available to the three groups of respondents, they are confounded with subscriber status and were omitted from these analyses.

Table 4 identifies the standardized canonical function discriminant coefficients of the cable–no cable subgroup comparisons. Cable users are maximally different in background characteristics in terms of being younger, of lower educational levels, higher income, and greater acquisition of new technology items. So, in this analysis, education does discriminate well the cabled from noncabled household within the market studied, controlling for other background factors. Among the choice and viewing variables, using television viewing guides and heavier viewing amounts best separate cable from noncable respondents. The canonical correlation is .40 (p<.001), and this set of variables correctly classified two-thirds of the members of each of these two groups.

Discriminant analyses were computed separately for the background variable set and for the choice and viewing variables. The background variables yielded a canonical correlation of .31 and correctly classified 64% of the respondents; the viewing variables produced a canonical correlation of .29 and classified 61%

Table 4. Discriminant Analyses by Subscriber Group Pairings

Background Variables	No Cable/Cable	Basic/Pay
1. Education	− .41[a]	− .57
2. Income	.43	− .57
3. Age	− .22	− .16
4. # of Children	.08	.19
5. Innovative attitudes	− .01	.42
6. Innovative ownership	.41	.29
Viewing process/use		
1. Planned viewing	− .16	.20
2. Guide use	.50	.54
3. Orienting searches	.02	.37
4. Reevaluation, alone	.18	− .30
5. Reevaluation, group	.02	.09
6. Viewing time	.42	.00
Canonical correlation	.38	.40
Correct classification	69%	68%

[a]Standardized canonical correlation coefficient.

correctly. This indicates that these variable subsets, taken alone, do equally well in discriminating cable subscribers from noncable subscribers, but that the combined solution yields a higher correlation and better classification power.

Table 4 also presents the results for the basic cable vs. pay cable discrimination. The most powerful discriminants in this comparison show that the pay subscribers had less education, more income, and more innovative attitudes. Again, education emerges here, although not in the bivariate analyses. Pay subscribers also tended to own more technology items and to have more children. Factors which distinguished the pay subscribers' approach to watching television included more guide use, more extensive orienting searches, and less channel changing during and between shows. They also tended to put more planning in what they would watch. The canonical correlation coefficient for this set of variables was .40, and 68% of the respondents were correctly classified, 70% of the basic, and 66% of the pay subscribers. Again, separate discriminant analyses for the background and viewing process variables produced coefficients of .32 and .26, respectively. Thus, the combined results were substantially stronger.

Regression analyses. These analyses enable us to examine the viewing process variables and amount of viewing, as they would be predicted from the background variables and the other viewing/process variables. For each dependent variable, two models were tested: one included only background variables, i.e., subscriber status, age, gender, education, income, number of children at home, and the two innovativeness measures; the second included the choice-process variables other than the one being predicted, in addition to the background variables.[1] As with the discriminant analyses, channel repertoire was excluded as a predictor because it is confounded with subscriber status. Two sets of regression analyses were run: one defined subscriber status as a dummy variable differentiating cable and noncable respondents (n = 278), and the second defined subscriber status as a dummy variable differentiating pay cable and basic cable subscribers (n = 176).

Table 5 contains the regression results wherein having or not having cable constituted the subscriber status differentiation. One overall comment is that the full model predictors do better than the set of background variables alone; although the table contains both sets of information, for our purposes here we will describe only the full model results. That the viewing-process model variables add predictive power is some empirical justification for the conceptual development taking place in these chapters.

Viewing time, planned viewing and guide use all have multiple Rs in the .33–.37 range in the cable–no cable comparison. Having cable continues as a significant predictor of how much time is spent watching television and how often a guide is used. For all three of these, the age of the viewer is important—but it is the younger viewer who uses a guide more often, and the older who does more overall viewing and more planned viewing. Viewing time is also predicted by the extent to which television is watched with children, and relates negatively with the education attained. Guide use also is associated with planned viewing.

The remaining dependent variables are better predicted, falling in the multiple R range of .57 to .68. More extensive orienting searches are conducted more regularly by younger viewers, with higher incomes. They are less likely to plan their viewing and more likely to change channels more often, both when viewing alone and when viewing with others. And these latter two forms of reevaluation co-occur, together with planned viewing and orienting searches, but with no strong associations with any of the background variables, save that males do more channel changing when viewing alone.

The final dependent variable—channel repertoire—is a function primarily of having cable, as previously noted as a basis for not including it as a predictor for the other variables. Guide use also is a significant predictor of channel repertoire.

The second regression analysis was based on substitution of the pay-basic dummy variable in place of cable/nocable (Table 6). That distinction serves as an important predictor only of guide use, within this set of variables—more guide use being predicted best by the pay-cable criterion. Amount of television viewing and the amount of planned viewing are both responsive to the age of the respondents, with older viewers doing more of each of these.

As with the first regression analysis, orienting searching, and changing channels when viewing alone and when viewing with others, are a package of interre-

---

[1] Control variables were presence/absence of remote control, proportion of time viewing alone, extent of viewing with children, and the frequency of the respondent's participation in program selection.

Table 5. Multiple Regressions Predicting TV Behaviors: Cable–No Cable Pairing

Dependent Variables	Demographics			Full Model		
	R	p	Predictors	R	p	Predictors
Viewing time	.31	.00	.17 cable	.35	.00	.20 age
						.18 cable
						.16 watch w/kids
						.14 education
Planned viewing	.28	.00	.22 age	.33	.00	.22 age
			− .11 sex			.15 selection role
						− .13 sex
Guide use	.28	.00	.20 cable	.37	.00	.19 cable
						.19 planned viewing
						− .14 age
Orienting search	.43	.00	− .40 age	.69	.00	− .31 plan viewing
			.12 income			.25 group reeval.
						.25 alone reeval.
						− .18 age
						.11 income
Alone reevaluation	.34	.00	− .26 age	.57	.00	.33 orienting
			.17 sex			.27 group reeval.
						.18 sex
						.13 watching alone
						.12 plan viewing
Group reevaluation	.34	.00	− .30 age	.61	.00	.30 orienting
						.25 alone reeval.
						− .16 selection role
						− .12 plan viewing
Channel repertoire	.64	.00	.58 cable	.68	.00	.54 cable
			.13 children			.11 guide use

lated behaviors, each predictive of the other. Searching is also predictable from the amount of unplanned viewing reported, and occurs primarily among younger viewers. Changing channels when viewing alone is further predicted from the amount of solo viewing done, and occurs more often among males and those of less education. Group channel changing is accompanied by more unplanned viewing, and is more likely to involve female viewers. Channel repertoire, interestingly, is predictable from having pay cable and having children; placing it in the fuller model attenuates its predictability.

   Results addendum.   Given a 35-channel system, we also determined the nature of channel clusters, i.e., does watching any one channel in such a system go along with watching any other channels, and which ones? Is there anything about how they are organized, e.g., their sequencing or their physical location, that directs audience traffic through them? Solely for exploratory purposes, then, we took the viewing data collected for each of the 35 channels in the system. The first dozen (the top bank on the three-bank push button channel selectors available in a majority of the subscribing homes) contained six network affiliates, one PBS station, one independent commercial station, HBO, MTV, CBC, and a lo-

Table 6.   Multiple Regressions Predicting TV Behaviors: Basic–Pay Pairing

Dependent Variables	Demographics			Full Model		
	R	p	Predictors	R	p	Predictors
Viewing time	.35	.02	.16 age	.39	.01	.17 age
Planned viewing	.25	.37	.19 age	.30	.36	.19 age
Guide use	.31	.08	.20 pay	.39	.06	.19 pay cable
Orienting search	.38	.00	−.31 age .16 income	.64	.00	−.35 unplanned .20 group reeval. .18 alone reeval. −.15 age
Alone reevaluation	.36	.01	−.22 age .19 income .17 sex	.57	.00	.33 group reeval. .21 sex .21 orienting .20 watching alone −.19 education
Group reevaluation	.30	.10	−.19 age	.60	.00	.22 orienting .32 alone reeval. −.19 plan viewing −.14 selection role −.13 sex
Channel repertoire	.34	.03	.20 children .17 pay	.39	.09	

cal programs channel. The second dozen included five satellite channels, five news and weather channels, a remote PBS station, and Cinemax. The last set of channels included three access channels, four educational channels, two government channels, the local library, and a channel for inspirational programming. Respondents answered about each channel whether they viewed it daily, weekly, monthly, or less, and their responses were weighted as indicated earlier to derive a days-per-month viewing measure. In terms of sheer viewing, maximum viewing went to the local network affiliates, each averaging 17 days per month. At the other extreme, among the last 11 channels, none was watched as much as 1 day per month.

An intercorrelation matrix of viewing of all nonpay stations by cabled households was factor analyzed, and five factors emerged. All stations had factor loadings of .38 or higher on a factor, save one—the local PBS station. The largest factor contained the last 11 channels (26–36), all viewed as little as just stated. Three other channels loaded most highly on this factor and also were minimally viewed, at 1 or 2 days a month; they were a nonlocal PBS station, a local programming channel, and a shared-time channel. This, we have labeled the *bottom bank* factor.

A second cluster comprised a *specialty cable* factor. It contained six channels—MTV, ESPN, CNN, WTBS, and sports news and business channels. A third grouping consisted of distant channels. Here were the three nonlocal network affiliates, a nonlocal independent channel, a Canadian station, and one shared-time satellite channel.

A fourth grouping was that of automated text news channels, i.e., world news, Michigan news, and weather, all character-generated channels. The final factor was that of the three local network affiliates.

These clusters for this single cable system accounted for 49% of the total variance in viewing. There also appears to be logic in their constituency.

DISCUSSION

This chapter began by identifying consistent and inconsistent findings in studies which attempted to isolate differences according to the status of cable subscribers. The information presented here has reviewed related literature and tested a series of simple relationships and in multivariate forms to attempt to determine the relevance of subscriber status. But in doing so, it also has introduced viewing and viewing-process variables into the mix, perhaps to add some substance to rather static demographic profiles. However, we have not yet addressed some major weaknesses alluded to in the opening pages. Most important, it seems to us, is the need for a cross-market analysis which includes some attributes heretofore ignored, e.g., the broadcast media environment in which cable systems are operating (cf. Baldwin, Abel, & Ducey, 1984); cable system attributes, such as the age of the system, and some indication of the changes that have occurred in the system since its inception; and even the presence/absence of different remote control/channel selector devices. Such a study would permit isolation of the effects of the number and kinds of cable channels available, the number of broadcast channels, cost(s) of the cable service, the relevance of remote control, etc. Perhaps much of this could be accommodated in a single national study with an adequately stratified sample that would address individual viewer issues; now that cable penetration is bordering 50%, such an attempt would not be overly wasteful, even without a national roster of subscribers. System questions would require a different approach.

Having reconfirmed a series of propositions about the demographic attributes of different levels of cable subscribers for the most part, it is clear that a better job has been done of differentiating subscribers from nonsubscribers than basic from pay subscribers. Perhaps the latter is truly a more fragile difference, one which is subject to promotions, to specials, to income problems—to more momentary rather than durable attributes. Perhaps the current difficulties of the pay networks in expanding their subscriber roster and/or maintaining their present levels is a function of this fragility, as manifested in the attractiveness of alternative entertainment means, e.g., videocassette recorders. Besides, it is considerably harder to relinquish basic than to relinquish a pay channel. And if the difference is not a function of television fandom—this study finds that basic and pay subscribers spend the same amount of time with television—then it likely is a content difference in part, and that has not been examined here, nor elsewhere to any considerable extent.

One other finding from this study bears on this pay–basic distinction, and that is what may be called the greater sampling viewing of the pay subscriber. Although the estimated viewing time is the same, the number of channels viewed within that time frame is substantially different; i.e., pay viewers spend less time with any given channel on the average, and the difference in channels viewed is not accounted for by the additional pay channel subscribed to. Further, this comes out again as a viewing attribute in the evidence that pay subscribers do more searching around for what to watch, and their increased use of a guide may make them more familiar with more channels. This argument would be even better supported if we found them also more frenetic while watching, i.e., engaging in more frequent channel changing behaviors; but alas, we do not yet find that correlative behavior. In fact, the discriminant results suggest the opposite pattern while watching—less changing. Indeed, that makes sense if they are watching the pay channels and if a primary motive for changing is to avoid commercials. At any rate, it may be more fruitful to look for viewing choice and viewing process differences as a means of differentiating pay from basic subscribers, than for additional background characteristics.

And this, of course, brings us straight into the issue of subscriber or viewer typologies created, not from subscriber status, but from the possible blending of viewing process variables. Here, they have begun to emerge at least as correlates of different kinds of viewer statuses. Leaving aside for the moment obvious questions related to the stability of these behaviors, e.g., how changeable are guide use tendencies or how reliable are self-reports of what otherwise might be electronically assessed, e.g., channel-switching behaviors, we can extricate from this set of findings at least two interesting issues to pursue.

First, the viewing-process variables have been demonstrated to be a set of somewhat interrelated behaviors, at least guide use, looking around for programs, and channel-changing at different points in a show. Treating them in this chapter as both predictor and predicted variables for analytic purposes blurs any notion of sequence or causality, and we leave much of that for another chapter in this volume. However, here they are treated as process variables and not related at all to content, and that is the first supplemental issue worth addressing. Do we not view different types or forms of content differently? Do we watch a football game in the same fashion as we watch the news or a soap opera? is the operational form of the question. How enduring are our viewing habits, or how triggered are they by alternative kinds or forms of content? More difficult still, how did we come to learn them and perform them in the fashion we do?

A personal example may help. When one of us watches a college football game on Saturday afternoon (assuming no home game at our own institution), it normally competes with at least two other football games on other broadcast or cable channels. Rarely would he sit without considerable channel-checking of the other games to see what's happening, if a key play is being replayed. And that would be considerably more channel-hopping than would be the case in watching an hour-long mystery in the evening. Further, it points out the confounding of type and form of content; the typical sports event has so much "down time" in it that it is ideal for him to exercise his quickening abilities at channel changing; he might be hard pressed to be so motivated or able while watching the network news. Yet that might be more frustrating, because he knows that there are at least two directly competing news broadcasts concurrently available. At any rate, it seems that a logical move from this content-free presentation of viewing-process behaviors would be to link them conceptually with content variables. At the same time, the issue requires examination of the extent to which these viewing postures persist across content variables and are learned behaviors, perhaps reflective of personality types as well.

A second issue goes back to our cable orientation and focuses on the findings about the number of different channels selected when three or more dozen are available. Here, how many is less interesting than which ones. It is apparent that most of the cable channels go virtually unused the vast majority of the time—the bottom-bank factor—which economically may be considered the pits by the franchise holder. One obvious influence is channel placement. If the current bottom bank of channels were distributed differently across the channels, would the factor still emerge? Another issue is what combination of channels and how many combinations of channels can conceivably accommodate the vast majority of the viewers' interests? Or how do media-mix models apply to the selection and offering of channel mixes in a cable system? What we have dubbed channel repertoire (the number of channels regularly used) has not been well explicated into what the composition of these repertoires might be. Given three individuals who regularly (weekly) watch programming on a dozen channels, how much channel overlap is there? Surely, if they are pay subscribers, their pay channel and the three network affiliates likely account for the vast majority of their viewing; how many more channels, and which ones, would satisfy them sufficiently to warrant retention of their cable subscription? This can of course be extended into an argument that would advocate reduction or elimination of most of the "minority" channels available, and perhaps so extended would be offensive to most of us. But without empirical support for the kind, composition, and quantity of individual channel repertoires, we are left with the depressing individual viewing statistics that show trivial use (less than weekly) of bundles of channels. Perhaps an equally promising approach would be the development of household repertoires. Given a sample of four-person families (two parents and two children), how similar or variable are the channel repertoires in such a grouping? How many channels would be needed in that family's cable system? Finally, how often does another channel drop into or out of such mixes?

The general recommendation here, of course, is to merge viewing style information with demographic and other background attributes as a sounder basis for differentiating cable users and nonusers. As penetration continues to increase, it is likely that demographics will be increasingly unsatisfactory as an explanatory system, and certainly little basis for maintaining subscribers.

## CHAPTER 17

The Playboy Profile and Other Pay Channel Subscribers

Bradley S. Greenberg, Sherri Sipes, and Julie McDonough

Among the more than three dozen cable audience studies published in the last 8 years, only one focuses on the distinction between the market for a particular premium cable channel and basic cable television (Ducey, Krugman, & Eckrich, 1983). Even that one does so laterally; its primary interest was in comparing basic and pay subscribers in a market where the only pay channel was HBO. It arrived at a profile of a particular premium channel which it treated as representative of pay channels. Other studies comparing pay, basic, and nonsubscribers to cable do so without differentiating which pay channel is involved, and, as often as not, whether it is one pay or multipay households being compared.

Why be concerned with differentiating among pay channel subscribers? For one, the content is not similar among HBO, Disney, and Showtime. Even moving across more similar premium efforts as HBO, The Movie Channel, and Cinemax does not mandate subscriber similarities. And if one is interested in the possible impact of content available through these channels, then comparability or the lack of it across subscriber segments is of much importance. Premium channels divert network audiences and warrant examination in their own right.

Among the several efforts to successfully market an "adult" channel, a 1980s euphemism for sex-oriented television, only the Playboy Channel is prominent. As of June, 1986, its subscriber roster was 650,000. Its still-limited penetration is not only a function of marginal public acceptance but of cable systems cautious not to offend city councils, and of multiple-system operators who refuse to tolerate it in any of their franchises. Its movies, news, game shows, and comedy sketches are strong in visual and verbal sex content; nudity and profanity are commonplace: bondage and female homosexuality are not uncommon.

Who is it that subscribes to Playboy? Dirty old men sitting before the TV set in bathrobes and flashing at the set? Liberated couples seeking a new high? Teens setting alarm clocks for 3 a.m. viewing after their parents are asleep? Swinging singles looking for doubles?

Tempting as it may be to make bad puns, there are some serious issues. For this chapter we will not deal with the issue of potential censorship and the First Amendment. Research has documented the increase in sex content on broadcast television fare (Sprafkin and others, 1977, 1979, 1981). Other researchers have begun to document some negative outcomes from persistent exposure to filmed and televised sex content (Zillman & Bryant, 1983; Baran, 1976a, b; Buerkel-Rothfuss & Mayes, 1981). In sum, they have found such diverse outcomes as stronger degradation of women, less personal satisfaction with sexual activity by contrast with televised models, and exaggerated estimates of real-world sexual activity.

Within this context, then, what might be expected of those who choose to pay a monthly stipend specifically to receive sexual content? According to Playboy research,* their subscribers have the following attributes, among others:

- two-thirds of the subscribers are married
- two-fifths have children under 18 in the household
- one-third are college graduates
- the majority are over 35, and one-fifth are over 50.

Let us speculate, then, on what might be associated with subscribing to the Playboy Channel. First, we would argue that content is selected to *reinforce* existing interests, that sex content on this medium is to supplement content available through other media, and less so to compensate for any presumed absence. We would posit that having the Playboy Channel is related to:

1. going to R-rated movies;
2. reading "adult" magazines; and
3. watching more television, with its increasing array of sex portrayals and discussion.

Second, it seems reasonable that viewers of the Playboy Channel would also be reflecting more *sexually permissive* attitudes, that

4. they are more accepting of sexual content on television in general.

At the same time that more permissive attitudes are anticipated with regard to the presentation of sexual content, it seems that the Playboy Channel does not provide parallel representations of men and women. Casual and unsystematic examination of the movies typically presented on that channel indicate that there is an emphasis on the subjugation and subordination of women. A working hypothesis may be that:

---

* Subscriber information was obtained from Andy Kline, research director at The Playboy Channel by telephone, in November 1984.

5.   Playboy Channel subscribers have more *sexist* attitudes; e.g., women are inferior to men, women prefer being dominated, etc.

And as modest outcome expectations from repeated exposure to programming in which nonmarital sex and domination of females are normative, then straightforward *social learning* theory would suggest:

6.   Viewers have different expectations about the occurrence of real-life sex, both before and during marriage, among both men and women.

It is also possible that Playboy subscribers seek different *gratifications* from this media experience than those who choose Disney or HBO. For example, it may be more so for vicarious excitement that such content is sought; on the other hand, if one argues that gratifications sought are as much generic to the individual as they are a function of any specific kind of content, it becomes more difficult to extract particular motives for watching this content. For this reason, a set of gratifications is examined in this study, but without known expectations.

Even with Playboy's own subscriber descriptions, we would expect the subscribers to be younger, with smaller households, and fewer children than would be found with some competitive, family-oriented premium channels.

METHODS

Field work consisted of a telephone survey of subscribers within a single franchise of a multiple-system cable television operator.

Sample.   The local cable franchise manager provided the research group with a computer printout of all current subscribers, as of 1 week before the field work began. The printout contained addresses, phone numbers, and what was being subscribed to. Samples of 150 were drawn from the following subscriber groups: Basic only, HBO only, Cinemax, Disney, and Playboy. Among the last three groups, the channel subscription was normally in addition to HBO (more than 90%). Only a handful in the sample subscribed to more than two channels, but it was not possible to construct samples in which Cinemax, Disney, or Playboy were the only premium channel taken. Of the 742 numbers in the original sample, 82 (11%) were lost because the numbers were disconnected or business numbers, or the citizen claimed not to have cable any longer. The completion rate on the remainder was 76%, with 16% refusals and 8% never contacted after at least three attempts. The completion rate was equivalent across subscriber groups, with n's ranging from 95 to 100 for each of the five groups. However, the correspondence between the system-provided listing of subscriber information and what the subscriber reported to the interviewer was far from isomorphic. Table 1 shows the lack of symmetry between the two information sources. Of particular note is the fact that barely half of the identified Playboy Channel sub-scribers reported that they received that channel, despite company information to the contrary. If all these were valid changes during the 3 weeks of field work, they represent 3% disconnects, 16% upgrade from basic, 10% downgrade and 12% upgrade for HBO, and 35%–47% churn for the Playboy, Cinemax, and Disney channels. For purposes of this analysis, respondents were considered subscribers to the channels they reported having. Thus, the n's used are those reported in the right hand column of Table 1.

Variables. From prior studies analyzing cable and noncable households, and pay vs. basic households (without consideration for number of pay channels or which ones they were), *demographic variables* were selected which tended to discriminate. These included: household size, number of children, income, edu-cation, age, and number of working persons in the household. In addition, gen-der, ethnicity, and marital status were obtained, but not used in the analysis: Gender was deliberately stratified in the interviewing process to yield half males and half females; there were too few non-Whites for analysis; marital status was evenly divided.

Media access variables included available television sets and the number con-nected to cable. *Media use* variables included self-reports of daily television hours, weekend television hours, non-TV movies seen the last month, a roster of specified R-rated movies seen, and household receipt of adult magazines like *Playboy* and *Playgirl*. To assess media sex content attitudes, respondents were asked the extent to which they believed sexual content on television is offensive, should be restricted for minors, and should be restricted for everyone.

The measure of *sexism* contained a series of four statements (a five-point scale of agreement/disagreement) dealing with gender role orientations, e.g., that men will always be the breadwinners, that most progress has been made by men and will continue that way. Responses to the individual items were summed to create the index.

Nine items expressing possible *gratifications* from television viewing were used: Watching TV to pass the time away, to get away from what I'm doing,

Table 1. Subscriber Groups by Self-Report and Company List

	According to Subscriber List:					
Said they had:	Basic only (n = 99)	HBO only (n = 95)	Playboy (n = 95)	Cinemax (n = 98)	Disney (n = 100)	(n)
Basic only	83	10	14	7	6	(120)
HBO only	12	73	22	15	10	(141)
Other	4	12	9	11	10	—
Playboy	*	*	50	*	*	(70)
Cinemax	*	*	*	65	*	(79)
Disney	*	*	*	*	65	(70)

*subsumed under other

when I'm bored, when there is no one to talk to, so I can learn about what's going on in the world, because it relaxes me, because it's exciting, because I just like to watch, and so I can learn about what could happen to me. Response categories were agree, not sure and disagree.

Finally, 12 items were used to assess *sex behavior* beliefs. Four were masking items, unrelated to content perceptions of the Playboy Channel. The remainder asked respondents to estimate how many times out of 100 each situation prevailed. For example, "Of every 100 men, how many are likely to have sex before marriage?" The same stem was given for the following: women/sex before marriage; married men/affairs; married women/affairs; women/rape victims; women/like being dominated by a man; men/like being dominated by a woman; and, women/like being seduced.

RESULTS

Analyses were done first to determine the extent to which the four groups of premium channel subscribers differed. Then, a discriminant analysis attempted to identify clusters of variables that might segment the different sets of premium channel subscribers.

Table 2 provides comparisons in terms of demographic attributes. Differences were obtained for household size, number of children under 18 years of age, and the number of working persons in the household. For household size, the Disney subscribers had the largest average size (nearly four persons per household), the other three groups did not differ (three persons per household). A similar pattern was found for having children under 18 years of age. Disney households averaged nearly two children, with the Playboy households significantly smaller than both Disney and Cinemax households. Playboy-subscribing households had the fewest number of working persons, significantly less so than HBO and Cinemax homes, but equivalent to Disney homes. No strong differences were found in terms of age, education, income, or marital status.

Table 3 examines media access and use variables, in addition to beliefs among

Table 2. Premium Channels by Demographics

Demographics		Playboy	Disney	Cinemax	HBO only	(p)[1]
a.	Household size	2.8	3.8	2.9	3.0	<.001
b.	Children < 18	.3	1.7	.4	.6	<.001
c.	Education	16.2	16.4	16.2	16.0	n.s.
d.	Age	31.8	33.3	30.9	32.1	n.s.
e.	Employed	1.3	1.5	1.7	1.7	<.02
f.	Income	3.0	3.1	2.8	3.0	n.s.

[1]The value is based on the results of univariate analyses of variance for each variable. Subsequent contrasts-tests among sub-group means are reported in the text.

subscriber groups about the presence of sex on television. Access is consistent across the groups: they have equivalent numbers of television sets and sets that are connected to cable TV. On all use and attitude variables, however, significant differences emerge. Playboy subscribers report watching more television on weekdays than any other group, and significantly more so than Disney or HBO subscribers. They also report watching more television on weekends, specifically more so than Disney subscribers. The two measures of exposure to sex-oriented media—R-rated movies outside the home, and adult magazines like *Playboy* and *Playgirl*, also set the Playboy subscribers apart. They claim to see the most R-rated movies, significantly more than Disney or Cinemax subscribers. They also are most likely to get *Playboy* or *Playgirl*; 40% of the Playboy Channel subscribers receive such a magazine, compared to 20% for Cinemax, 10% for HBO, and 0% for Disney subscribers.

Respondents were asked whether sex on television was offensive and whether it ought to be restricted for everyone. Playboy subscribers were least likely to advocate such a stance and Disney subscribers were most likely.

Table 4 presents the attitudes of the subscriber groups toward gender roles, and their beliefs about the occurrence of selected real-life sexual behaviors. The measure of sexism consisted of four items; the possible range of scores was 4–20, with a low score reflecting less sexism in response to items that stated men would continue to be the basic breadwinners, father should have the final authority over children, and that most progress would continue to be made by men. No significant difference emerged, although the direction of the means shows stronger sexist attitudes among Disney subscribers. Playboy subscribers were not more or less sexist than any other premium-channel group.

Beliefs about the frequency of occurrence of premarital sex among men and women, extramarital sex, sexual domination, and seduction are also tabled. Three of eight items show expected differences; Playboy subscribers were more likely to believe that more men *and* more women have sex before marriage than the other subscriber groups, and significantly more so than the HBO and Disney

Table 3.   Premium Channels by Media Access, Use, and Attitudes

		Playboy	Disney	Cinemax	HBO	(p)
Access						
a.	# of TV sets	1.8	1.8	1.7	1.9	n.s.
b.	Sets w/cable	1.3	1.2	1.2	1.3	n.s.
Use						
a.	TV hours/weekdays	3.8	3.3	3.6	3.0	<.03
b.	TV hours/weekends	4.5	3.8	4.1	4.1	<.06
c.	R-rated movies	2.6	1.9	2.0	2.3	<.03
d.	Adult magazines	.4	0	.2	.1	<.001
Attitudes						
	Restrict TV sex	3.9	5.5	4.7	5.0	<.001

Table 4.   Premium Channels by Beliefs About Sex Roles and Behaviors

	Playboy	Disney	Cinemax	HBO	(p)
Sex Roles					
Sexism scale	8.9	9.7	9.0	9.0	n.s.
Sex Behaviors*					
a. Men/sex before marriage	88	76	85	82	<.002
b. Married men/affairs	43	43	43	43	n.s.
c. Women/sex before marriage	71	56	68	62	<.001
d. Married women/affairs	36	33	33	36	n.s.
e. Women/rape victims	17	17	19	19	n.s.
f. Women/like being dominated by men	36	31	36	29	<.04
g. Women/like being seduced	53	51	60	50	n.s.
h. Men/like being dominated by women	20	10	17	22	n.s.

*Means are responses to this question format: "Of every 100 _____, how many _____?"

groups. The same pattern appears for the third item of note; Playboy subscribers believe that there are more women who like being dominated by men. For all three of the differentiating items, it should be noted that the Cinemax subscribers had the same pattern as the Playboy Channel subscribers. There were no consistent differences in the frequency with which these subscribers believed that married men and women have affairs, that women like being seduced, that women are rape victims, or that men like being dominated.

For none of the nine gratifications items, e.g., I watch TV because it is exciting, to get away from what I'm doing, when I'm bored, were consistent differences found, and these results are not tabled. For "It passes the time away," Playboy subscribers were more endorsing than any other group.

In review, then, of the tentative hypotheses we formulated, support was found for these propositions:

• Playboy Channel subscribers reinforced their interests in adult TV fare by going out to see more R-rated movies and by subscribing to a sex-oriented magazine.
• Playboy households were smaller and had fewer young children (particularly in contrast to large Disney families).
• Playboy subscribers professed several beliefs about the real-life occurrence of sexual activity that were significantly stronger than those of the other subscribing groups. This was confined to beliefs about premarital sex and male domination of women.

No support was found for any differences in gratifications sought from the Playboy Channel, nor for the notion that Playboy Channel subscribers would be more (or less) sexist.

Discriminant analysis. The full set of variables in the prior analyses, plus some not tabled, were submitted to a discriminant analysis. This procedure groups the variables into "functions" and then maps the criterion groups on those functions, to the extent they fit. Table 5 presents results from this analysis. Two of the three functions obtained were statistically significant; all three are presented, for reasons discussed below.

The first function appears to be that of the *Traditional Family*. The number of children under 18 predominates, and is clustered with lower income and the belief that a successful marriage *and* a successful career do not comingle for women. Also included on this function is the practice of going out to movies more often, and total family size. Disney subscribers are located alone on the extreme axis of this function; all other premium channel groups are located somewhat beyond the midpoint of this function, with the Playboy subscribers being the most discrepant.

The second function appears to be that of *Sexism*, with the beliefs that men do *not* like to be dominated by women and that women like to be dominated by men linked to a heavy dose of daily television viewing. Also clustered here are the beliefs that more women have sex before marriage, but that fewer married

Table 5.   Discriminant Analysis Among Premium Channel Subscribers

Function	Eigenvalue	Canonical Correlation	Wilks' Lambda	Chi Square	df	(p)
1	.552	.597	.455	225.46	114	<.001
2	.257	.452	.706	99.67	74	<.02
3	.127	.336	.887	34.31	36	.55

Function 1*		Function 2		Function 3	
(.94)	# of children	(−.64)	Men dominated	(−.55)	Sex magazines
(−.35)	Combine career/ marriage	(.60)	Women dominated	(.47)	Restrict TV sex
(−.23)	Income	(.56)	Daily TV	(.45)	Women like being seduced
		(−.45)	Weekend TV	(.34)	Sets cabled
		(−.33)	R-movies	(.32)	TV when bored
		(.29)	Women/sex before marriage	(−.31)	Marriage for love
		(−.24)	Married women and affairs	(−.26)	Sexism scale
		(.20)	TV exciting	(−.21)	TV to relax
				(−.20)	TV to learn

Group Centroids			
	F1	F2	F3
HBO	−.21	−.52	.24
Cinemax	−.26	.71	.16
Disney	1.48	.09	.31
Playboy	−.55	.16	−.94

*Variables listed are those with rotated standardized discriminant coefficients ≥.20.

women have affairs. Television is watched because it is exciting, but fewer R-rated movies have been seen. Cinemax subscribers are best described by these variables, and HBO subscribers least well fit these characteristics. The other two subscriber groups fall near center on this function.

The third function, although not statistically significant, warrants attention because it best describes the *opposite* of the Playboy subscriber. Considered an *Anti-Sex Content* function, the clustered variables with the strongest loadings include less exposure to sex-oriented magazines, considerably more restrictive attitudes toward the presentation of sex on television, that TV viewing is done when you're bored, and not to relax or to learn, and a negative weighting on the sexism scale. One contradiction among the clustered variables is the stronger belief that women like being seduced; that fits an interpretation of this function as a "conservative" one, but does not meld with the negative loading of the sexism scale. Although none of the subscriber groups examined is very strong on the positive end of this function, the Playboy subscriber groups is quite strong on the negative end; these are attitudes and behaviors they do *not* accept. The reader is reminded that this third function was not significant, but nevertheless intriguing.

This analytic mode also permits the groups to be classified in a matrix crosstabulating actual group membership with predicted group membership based on the functional scheme. By chance, 40% of all the cases would have been classified correctly; actually, 60% were correctly classified (Press $Q = 65.28$, $df = 3$, $p < .001$). By subscriber group, these results were obtained:

Subscriber Group	Correct by Chance	Actually Correct
Playboy	20%	49%
Disney	17%	55%
Cinemax	23%	43%
HBO	40%	77%

DISCUSSION

There is no surprise in finding that subscribers to a particular premium channel, with a specified kind of content, are different from those who do not choose that content, or different from those who have chosen alternative premium channels. Finding out exactly how they differ is informative; it emphasizes the need in cable-related research to make this kind of distinction, and not to conveniently and artifactually lump all pay subscribers into a single grouping.

In designing the study specifically to look at the potential impact of exposure to the Playboy Channel, we did not give sufficient attention to the kinds of impact that might accrue from the alternative premium channels. For example, a parallel examination of Disney subscribers, in terms of the potential prosocial outcomes of persistent viewing of films and other programs emphasizing altruism, sharing, affection, problem-solving, etc. seems warranted if one is focusing on social effects research. Where content is relatively consistent across nonpay channels, e.g., MTV, Nickelodeon, CNN, the same questions emerge.

Within the Playboy context, the study taps the surface of outcome measures of interest. It did not assess attitudes toward sexual activity, only toward its presence on television, and it did not get into personal behaviors and preferences. For example, is there reason to believe that viewing such content increases or decreases marital (couple?) satisfaction with sexual experiences. Baran's research (1976a,b) with teenagers suggests that they are more disappointed; would that hold with an older, presumably more experienced, set of viewers? Given the lack of differences for the gratification measures used, we have little understanding of why that channel is sought; measures of gratifications that are channel-specific should be developed. More importantly, perhaps an overtime study would yield more inclusive information as to what viewers derive from this programming. The kinds of correlational findings obtained here, that Playboy viewers are more tolerant of sex content on television, do not permit us to differentiate reinforcement of viewing patterns from reinforcement of attitudes.

More systematic analysis of the content of Playboy Channel programming is needed. Best guesses have been used here about what the themes emphasize. Although coding schemes have been developed for sexual content on broadcast television, they are not likely to be adequate for the programming on this channel. Nevertheless, that initial step was omitted here. That Playboy viewers were more likely to project the occurrence of premarital sex, but no more likely to project extramarital sex, may or may not correspond to the content emphases in the programming.

The discriminant analysis has been used for exploratory purposes. It suggests that a variety of life style values and behaviors can clarify premium channel orientations. Certainly, it suggests marketing strategies for the channels other than HBO (given HBO's universal distribution in this sample). The traditional values of the Disney fan might well have been anticipated; and with larger families and less per capita income, further penetration may require alternative pricing strategies. An alternative multivariate procedure, multiple regression analysis, would assist in determining the extent to which these variables account for subscribership of each of the premium channels.

This study has not laid out the full range of groups of interest in such research, beginning with nonsubscribers, moving to basic cable subscribers, and subsequently to each of the premium-channel options available in a given franchise area. Nor has any other study. Perhaps the Playboy subscriber is no different than those without cable in terms of attitudes, beliefs, and practices. Is there anything that cumulates with this progression, from nonsubscribers to multicable, multipay subscribers, beyond the sheer number of options available to watch on television?

One serendipitous finding worth comment is the relative "churn" in Playboy Channel subscribers within the 2 weeks between the time we received a current subscriber list from the cable firm and the completed interviewing of the sample. Playboy subscribers may indeed downgrade at a faster rate. Or the cable firm's subscriber list may be quite inaccurate. On the other hand, substantially more subscribers of that channel may not want it known that they are receiving it. Had this been anticipated, a larger sample of Playboy subscribers would have been created; then, those who acknowledged their current Playboy subscribership would have been compared with those who said they did not receive it (but who were receiving it, according to company billing records). The question of who "fibs" about getting the Playboy Channel could then have been investigated.

Perhaps we have acquired as much information about what Playboy subscribers are not, as about what they are. The discriminant analysis locates them on opposite poles for both the traditional notions and the antisex content dimensions. Nationally, they remain a small number of households, by contrast for example, even with Disney (2.6 million) and Cinemax (3.7 million). Nonetheless, their affinity for sexual content makes them fertile targets for research questions of national interest.

CHAPTER 18

Music Video Viewers

Ronald Paugh

One cable-television vehicle which has enjoyed tremendous growth for the past few years is Warner Amex's MTV: Music Television. Since success usually breeds imitation, many programs with MTV's format have been introduced. A few examples include Night Flight, VH-1, and Friday Night Videos.

As marketers have profiled volume segments, i.e., nonusers, light vs. heavy users for certain product classes and brands (Twedt, 1964), so too have media researchers profiled the characteristics of user segments of different media. Advertisers continually seek information about the personal-characteristic composition of television-program audiences. Advertisers desire data, especially demographic and psychological information about program audiences, simply to understand them better, reach them more efficiently, and infer what copy might be most effective (Cannon & Merz, 1980; King & Summers, 1971). Since product and brand target markets are increasingly being defined in terms of psychographic variables, and different media and programming can be considered as products, this study concentrates on MTV viewer's life-style characteristics for classification and description.

Psychographic or life-style variables are combinations of personality, socio-economic, attitude, and behavioral constructs measured indirectly. When combined with demographic characteristics, a more complete and understandable profile of individuals is provided. Numerous researchers have used psychographics in profiling various media audiences in the past (Frank & Greenberg, 1979; Lumpkin & Darden, 1982; Peterson, 1972; Villani, 1975), but there has been little published research findings specific to the MTV audience.

For this study, initial items were selected from the Activities, Interests, and Opinions (AIO) library of Wells and Tigert (1971). This particular test battery was used because of its relative popularity and accessibility to researchers. In addition, the items have satisfactory reliability and validity scores (Wells, 1975). Other life-style and music-related media usage variables were also included in the analysis. Examples include media-usage behavior and music-related media-usage behavior. The primary industry concerns of MTV's influence on radio listening habits, and its influence on record and tape sales, guided the selection of the variables (Forkan, 1981). Table 1 lists the variables examined in this study.

METHOD

Data were obtained from a random sampling of 499 persons between the ages of 18 and 34 in the Lansing-East Lansing area. This area, located in central Michigan, is a top-100 television market. There are approximately 115,000 television households in the sample area. Although the 12–34 age category is MTV's key market demographic, the analysis was confined to the 18–34 age category. This was done because it was believed that the younger age category respondents

Table 1.   Variables Analyzed

Demographic Variables	Psychographic Variables*
Age	Religious
Sex[a]	Conservative
Marital Status[b]	Culture
Family Size	Materialistic
Income[c]	Hi-tech
Education[d]	

Life-Style Variables

A.   MTV Viewing Behavior
1.   MTV viewing per day
2.   MTV as background (%)
3.   Number of companions
4.   Watch other music shows

B.   Other Media Usage
1.   Television viewing behavior
2.   AM radio listening (minutes)
3.   FM radio listening (minutes)
4.   Cable TV subscriber (months)
5.   Time spent reading rock magazines (weekly)
6.   Movies attended last month

C.   Music Related Media Usage
1.   Stereo equipment ($ invested)
2.   Records/tapes purchased last month
3.   MTV influence on record/tape purchase
4.   Rock concerts attended last month

*See ''Method'' for operationalization of these variables.
[a]dummy variable: 0 = female; 1 = male
[b]dummy variable: 0 = not married, 1 = married
[c]dummy variable: 0 = less than $25,000 yearly; 1 = $25,000 or greater yearly
[d]categories: 1 = less than high school; 2 = some high school;
            3 = high school graduate; 4 = trade/vocational school;
            5 = some college; 6 = Bachelor's degree; 7 = graduate training

would have been unable to respond to the specific questions regarding stereo equipment and record/tape purchases, and the life-style questions regarding cultural and materialistic attitudes. Ninety-five percent of the sample have cable television available in their area, 80% subscribe to cable television, 91% are familiar with MTV, and 80% have watched MTV at least once. The usable sample size, however, was 303, because only those with cable television and those who spend some time on an average day watching MTV were considered for further analysis. Moreover, 81% of MTV viewers typically are in pay homes, thus necessitating this requirement (*Broadcasting*, 1984).

Trained interviewers (undergraduate research students) conducted telephone interviews from a central location with a research instrument that included 35 questions, divided into three specific sections (see Table 1). Over 40 psychographic items (four constructs with approximately 10 items per construct) were administered in a pretest questionnaire, using a five-point, strongly-agree to strongly-disagree response set. All initial items were checked using Cronbach's (1951) alpha, a method for determining the internal consistency of a scale with multipoint items. Items employed in the final version of the questionnaire had alpha scores above .65, acceptable scores for exploratory research.

Four psychographic value scales were created by adding the individual item scores and then determining the average. A principal components factor analysis, using the varimax rotation criterion for factor rotation, was performed on the items in each scale. These 40 items were thus reduced to 12 items in the analysis. The factor loadings (in parentheses) follow each of the statements, and hence, the name, or underlying dimension.

1. CONSTRUCT 1: RELIGIOUS
   a. I go to church regularly (.81)
   b. If more Americans were religious, this would be a better country (.51)
   c. I often read the Bible (.72)
2. CONSTRUCT 2: CONSERVATIVE
   a. Today, most people don't have enough discipline (.38)
   b. Our society makes children grow up too fast today (.70)
   c. Everything is changing too fast today (.67)
3. CONSTRUCT 3: CULTURE
   a. I enjoy going through an art gallery (.58)
   b. I enjoy going to the ballet, opera, or symphony (.79)
   c. Classical music is more interesting than popular music (.53)
4. CONSTRUCT 4: MATERIALISTIC
   a. I like to drive a powerful car (.47)
   b. I would like to own the most expensive things (.83)
   c. I would like to own all the latest gadgets (.75)

For the fifth construct, "hi-tech ownership," respondents were asked to respond *yes* (1) or *no* (2) to the following five questions:

1. Do you have television in your household?
2. Do you have any video games?
3. Do you have a home computer, such as the Apple, Commodore VIC-20, or TRS-80?
4. Do you have a videocassette recorder?
5. Do you have a videodisc player?

Each *yes* response was summed together to create an index of "hi-tech ownership." Since all respondents had cable television, question 1 is a constant, but it was a check on the validity of the sample.

ANALYSIS

The analysis performed involved four basic steps. First, the original set of demographic, psychographic, and life-style variables was screened in preliminary bivariate analyses to pinpoint those variables that individually do the most to distinguish between "heavy" viewers and "light" viewers of MTV (defined in "Discriminant Analysis" section). Seventeen of the 25 variables were found to have significant univariate F-ratios ($p<.10$) and were thus included in the subsequent analysis. This significance level was judged to be adequate, given the exploratory nature of the research. Although sex was not found to be statistically significant, it was retained in the discriminant analysis, as a key demographic variable. Knowledge of several simple variables—age, sex, educational attainment, for example, provides a reasonably accurate guide to the type of communication content a given individual will or will not select from available media and media types. This "social categories" approach is a descriptive formula that can serve as a basis for rough prediction and as a guide for research (DeFleur & Ball-Rokeach, 1982).

Second, a discriminant analysis was performed in order to identify demographic, psychographic, and life-style differences between heavy and light viewers of MTV. Third, the total usable sample was split into two samples, an analysis sample and a holdout sample. This was necessary to eliminate the upward bias in classification results which occurs when the same data used to generate the functions are used to test them (Frank, Massy, & Morrison, 1965; Morrison, 1969). No definite guidelines have been established for dividing the sample into analysis and holdout groups, but a 90–10 split has been recommended in the literature (Morrison, 1969). In this study, a 75–25 split between the analysis and holdout groups was implemented, owing to the exploratory nature of the study. Moreover, it was judged that a larger holdout sample would lend more validity to the discriminant function obtained. Fourth, the reliability of the results obtained in the third step was examined by attempting to predict MTV viewers' behavior type, i.e., light vs. heavy viewers, in the holdout sample by using the demographic and AIO information and the discriminant coefficients obtained in the analysis sample. The Press Q statistic was used to determine if the procedure discriminated better than chance (Press, 1972).

The analytical procedure used in this study is discriminant analysis, a multivariate procedure particularly appropriate when group membership is known. The basic purpose of discriminant analysis is to weigh and linearly combine the discriminating variables in the study, such that the resultant groups are as distinct as possible in terms of characteristics analyzed. The groups are defined by the particular research situation. In this study, the groups are "light" MTV viewers and "heavy" MTV viewers, as determined by the response to the question on "how many minutes on an average day do you spend watching MTV?" The frequency distribution to this question elicited a mean of 31 minutes and a modal response of 60 minutes. The response of 10 minutes divided the sample into two approximately equal groups. That is, approximately 50% of the sample viewed at least 10 minutes or more of MTV per day, and approximately 50% of the sample viewed less than 10 minutes of MTV per day. Those viewing 10 minutes or more of MTV per day were defined as the "heavy" viewer segment, and those watching less than 10 minutes per day the "light" viewer segment. Conceptually, because of the nature of MTV programming, that of 4- to 5-minute film sequences of rock groups performing either in concert or in a studio setting, it would not be likely that viewers would watch more of this if they were not interested in rock music or this particular type of programming format. Statistically, an equal or approximately equal split between two a priori defined groups obtains the clearest picture of determining how well the independent variables discriminate. Moreover, a sharp break occurred at the 10-minute viewing level, wherein: (a) the 1- to 9-minute viewing level was fairly evenly distributed, (b) a small proportion (15%) of the respondents viewed between 11 and 27 minutes, and (c) a much larger proportion (30%) of the respondents reported viewing between the 30- to 60-minute level.

RESULTS

Table 2 presents the mean comparisons between light and heavy MTV viewers. A profile of heavy MTV viewers emerges. Heavy viewers tended to be younger, less well educated, from larger households, and less likely to be married. Psychographically, heavy viewers were less culturally oriented, more materialistic, less conservative, and tend to make use of (own) more high technology equipment.

Heavy viewers were more likely to use MTV as background. They watched MTV with more companions, and estimated that MTV had a greater influence on their record/tape purchases. They also purchased more records in a month, and spent more time reading rock magazines. Heavy MTV viewers attended more

Table 2. Bivariate Significance-of-Difference Tests Between Heavy Viewers and Light Viewers of MTV

Variable	Heavy Viewer Means	Light Viewer Means	F	Significance
Demographic Variables				
Age	24.5	26.6	14.9	<.01
Sex (% Male)	55%	60%	.54	N.S.
Marital Status (% married)	39%	51%	4.0	<.05
Family Size	3.3	2.9	3.5	<.07
Income	.40	.43	.13	N.S.
Education	4.0	4.5	9.4	<.01
Psychographic Variables				
Religious[1]	2.7	2.8	.17	N.S.
Conservative[1]	3.5	3.7	4.4	<.05
Culture[1]	2.7	2.9	6.1	<.05
Materialistic[1]	3.1	2.7	14.7	<.01
Hi-tech[2]	1.5	1.3	2.7	<.09
Life-Style Variables				
A. MTV Viewing Behavior				
1. MTV Viewing per day (min)	46.1	4.8	79.8	<.01
2. MTV as background (%)	43%	29%	8.3	<.01
3. Number of companions watch MTV with	1.6	1.0	8.8	<.01
4. Watch other music shows (% yes)	63%	57%		N.S.
B. Other Media Usage				
1. Television viewing (min/day)	169.9	136.2	3.1	<.08
2. AM radio listening (minutes/avg. day)	6.0	13.0	2.03	N.S.
3. FM radio listening (minutes/avg. day)	172.0	180.0	.08	N.S.
4. Cable TV subscriber (mo)	31.8	36.9	2.6	<.10
5. Minutes/week spent reading rock magazines	70.0	19.3	3.2	<.08
6. Movies attended (monthly)	1.5	1.0	8.7	<.01
C. Music Related Media Usage				
1. Stereo equipment ($ invested)	$849	$703	1.06	N.S.
2. Records/tapes purchased (monthly)	1.7	.98	6.5	<.05
3. MTV influence on record/tape purchase (0–10 scale, 10 = highest degree of influence)	6.1	3.6	9.9	<.01
4. Rock concerts attended (monthly)	1.6	1.5	.14	N.S.

[1]Scale can range from 1 to 5; 5 = Strongly Agree, 4 = Agree, 3 = Neutral, 2 = Disagree, 1 = Strongly Disagree
[2]Scale can range from 0 to 5; 0 = No "high tech" ownership, 2.5 = moderate "high-tech" ownership, 5 = high "high tech" ownership

Table 3.   Summary Table—Discriminant Analysis
Between Light and Heavy Viewers of MTV

Eigenvalue	Canonical Correlation	Wilks' Lambda	Chi-Square	df	Significance
.21071	0.41718	0.82596	29.924	7	.0005

Standardized Discriminant Coefficients		Group Centroids	
Variable	Coefficient*	Light	Heavy
Materialistic	0.49975	−0.69585	0.29909
TV viewing (minutes)	0.32470		
Records purchased (monthly)	0.30611		
Sex	−0.28662		
Age	−0.27646		
Culture	−0.23919		
Family Size	0.20172		

*all coefficients are significant at p < .001

movies, watched more TV, and were newer subscribers to cable. Radio listening, gender, income, religious orientation, and rock concerts attended monthly were not different between the two groups.

Table 3 presents the results of the two-group discriminant analysis. Within this analysis, the basic conditions were met, wherein: (a) no perfectly correlated variables were used, and (b) Box's M statistic testing the null hypothesis of homogeneous group covariance matrices was significant at the .05 level.[1] Certain audience characteristics can be observed by looking at the extreme values of the discriminant coefficients for each variable. A variable contributes most to the probability of classification for which it is most positive. Conversely, negative coefficients indicate the extent to which high scores on a variable are not likely to be associated with the particular audience. Variables whose coefficients are near zero for any group do not much affect the probability for that group. In any event, the coefficients serve to indicate the relative contribution made by the respective independent variables in predicting the dependent variable. Upon examination of the coefficients, it can be seen that heavy MTV viewers are more materialistic, watch more television, purchase more records, tend to be male, are younger, are less oriented to cultural fare, and have a larger family size than light MTV viewers.

The amount of variance explained by the entire model is provided by the eigenvalue and associated canonical correlation. One can interpret the square of the canonical correlation as the proportion of variance in the discriminant function that is explained by the groups. Approximately 17% of the variance (.41718 squared) can be explained by the discriminant function.

Another criterion for evaluating discriminant functions is to examine the group centroids (means) of the two groups. The more disparate they are, the

---

[1] Box's M = 68.738, approximate F = 1.4129, df = 45,3052.2, p = .0396.

Table 4.   Confusion Matrix: Predicted Group Membership of Light and Heavy Viewers of MTV—Analysis Sample

Category	Chance Probability*	Predicted Viewer Segment	
		Light	Heavy
Light MTV viewer	46.2%	62.7%	37.3%
Heavy MTV viewer	53.8%	19.5%	80.5%

Percentage of "Grouped" Cases Correctly Classified: 75.2%

Chance (modal category) = 53.8%

$$\text{Press Q} = \frac{(169 - 127(2))^2}{169} = 42.75$$

df = 1

p is less than .001

*These percentages reflect the respective proportions of individuals in the study deemed "light" MTV viewers (140/303) and "heavy" MTV viewers (163/303).

stronger the separation of the two groups. It would appear, however, that a meaningful separation has been achieved, given the centroids obtained.

A final indicator of the ability of these variables to discriminate between the two viewing level groups is the confusion matrix of the analysis sample found in Table 4. The confusion matrix indicates how well the derived set of variables can predict group membership as compared to chance probability. As noted in Table 4, the study variables correctly predict 75.2% of group membership. The Press Q statistic, a chi-square test for discriminatory power with one degree of freedom, equals 42.75, significant at p < .001. Thus, the classification procedure does better than chance (assignment of all respondents into the modal category). However, since the same data are being used to evaluate the procedure as well as define it, there is an upward bias in the classification results. This bias is eliminated by use of a holdout sample. The holdout sample has been classified on the basis of the classification-function coefficients which were derived from the analysis sample. This procedure is used to validate the discriminant model. Table 5 presents the confusion matrix for the holdout sample. The procedure is significantly better than chance, $X^2 = 7.69$, df = 1, p < .01.

Table 5.   Confusion Matrix: Predicted Group Membership of Light and Heavy Viewers of MTV—Holdout Sample

Category	Chance Probability	Predicted Viewer Segment	
		Light	Heavy
Light MTV viewer	46.2%	50.0%	50.0%
Heavy MTV viewer	53.8%	20.6%	79.4%

Percentage of "Grouped" Cases Correctly Classified: 69.2%

Chance (modal category) = 53.8%

$$\text{Press Q} = \frac{(52 - 36(2))^2}{52} = 7.69$$

df = 1

p is less than .01

In respect to the demographic traits examined, several variables which have been found relevant in previous research were not found to be significant in this study. Most notably, income and education were not important characteristics. Although some personality constructs were significant as discriminators, religious and conservative orientations did not discriminate. The univariate test elicited a difference in the number of companions co-viewing MTV for the two groups, but was not found to be a significant variable in the multivariate analysis. It would seem likely that some situational characteristic is operative and needs to be explored further.

CONCLUSION

This study examined variables discriminating between light and heavy viewers of MTV. The results may help marketers, advertisers, and social scientists better understand who tends to be heavy and light viewers of this new type of programming format. This input could be very useful in placing advertising and determining creative copy platforms of commercials placed in this type of vehicle.

Further research is needed in areas related to the issues examined here. For example, why do avid fans of MTV watch for several hours at a time and what situational circumstances tend to enhance the MTV viewing experience? A taxonomy for situational characteristics (Belk, 1974, 1975) has not yet been delineated, but physical surroundings (i.e., weather), social surroundings (i.e., presence of other people), and antecedent states like mood have been researched. Of special interest is the finding that leisure activities (such as television viewing) seem to have the strongest ties to these situational characteristics (Hornik, 1982). Although the bivariate significance of difference test indicated heavy MTV viewers use MTV as a background element and view in larger groups, these variables did not appear in the discriminant function (except as larger groups may include larger HH). Further investigation of viewing situations may yield some insight into explaining the success of this new type of programming format.

Further research efforts should be directed to determine the types of television and other music shows watched by this audience segment. Radio formats most preferred and magazines read, especially rock magazines, should be analyzed. It may be an appropriate strategy for firms in the communication industry to diversify into those media which are consistently exposed to the same consumer groups. Thus, when the media offer their product (i.e., space, time) it can offer the advertiser a complete ''package''—a readily defined, reachable market segment across media.

Manipulating Viewing Through Field Experiments

## CHAPTER 19

Free Cable Service as an Incentive

Carol Mackey

INTRODUCTION

Cable marketing is designed to obtain subscribers, upgrade subscribers, and retain subscribers. Obtaining new subscribers is typically partialled into two tasks—signing them up when the system begins to operate, and later going after those who have said, "No, thank you." This project was targeted at the latter, at that set of nonsubscribers who have been most resistant to traditional marketing approaches. It gauged their receptiveness to a specific marketing strategy and their subsequent cable subscribership decision.

Typically, cable systems have thrived when begun in small- to medium-sized towns, deprived of much in the way of local television stations, or quality access to more remote off-air signals. Initial penetration has often been at the level of half of the homes passed by the cable. This has meant early profitability and perhaps an initial lack of regard for the remaining households passed but not subscribing. However, because of subsequent churn (disconnects or reduction in number of services obtained) and lack of continuing growth of new subscribers, and upgrades of existing services taken, franchise operators have refocused their attention on the nonsubscriber. Maintaining revenues has become as much of a concern as has increasing them. Marketing efforts that have best characterized the industry have included customer testimonials (*Cable Marketing*, August, 1984), door-to-door selling (*Cable Marketing*, January, 1985a), packaging various combinations of services (*Cable Marketing*, January, 1985b), free trials of individual channels, market segmentation (identifying different subgroups of nonsubscribers), and attempts to unify the company's image through a consolidated program of media advertising, program guides, direct mail, etc. (Russell, 1984).

Prior research on the nonsubscriber has largely been based on identifying demographic distinctions. The typical nonsubscriber is more likely to be older, have a lower income, not own a home, have a smaller family, be less of a TV fan, watch fewer channels, own fewer pieces of communication technology, e.g., VCRs and computers, and have a higher education than the typical subscriber (Agostino, 1980; Sparkes, 1983b).

A cable-industry-sponsored study indicated that the main reasons for subscribing to cable were improved television reception and greater access to movie and sports programming. Chief among factors in the likelihood of subscribing was the respondent's attitude toward television (Cable/Video Research Center, 1983).

One major multiple system operator combined demographic and psychographic data from its nonsubscriber markets to create a market segmentation plan of four nonsubscriber groups which were then targeted for different marketing strategies. First were the "seniors," generally older, single females with lower than average income and higher than average education. Market strategy here emphasized good reception, value, variety in news and information and quality programming. Second were the "up-scale nonsubscribers," usually married, older females with higher than average education and income. Marketing to them involved stressing high quality educational/intellectual programming, current events, good reception, and specific programming. Third were "young singles/couples," often without children, average or lower than average education and income, but heavy television viewers. The plan for this segment focused on recent movie offerings, 24-hour and (some) commercial-free programming, variety, and convenience. Fourth and finally were "families," of average to below average income and average education. They were told of children's programming, educational programming, value, something for everyone in the family, current events, and limited sex and violence. The MSO claims a high success rate for this segmentation plan aimed at nonsubscribers (Gilbert, 1983).

At another extreme is the marketing plan developed by Donnelly and Simmons with their ClusterPlus System. This plan involved 47 different demographic and lifestyle clusters, each presumably capable of being targeted for marketing by cable operators, depending on the operators' ability to match their nonsubscriber segments with the various clusters identified (Donnelly Marketing Information Services and Simmons Market Research Bureau, Inc., 1983).

The present evaluation, in three small town cable systems, gauged nonsubscribers' television viewing behavior and preferences, their perceptions of cable, their existing attitudes toward television and cable, and their reasons for not having cable, among other attributes. Furthermore, it incorporated a free trial cable offer marketing technique as a means of assessing the utility of that approach with nonsubscribers in those systems.

METHODS

This research project was designed to test a marketing program which involved free installation and 1 month of free cable television service to nonsubscribers in three cable systems owned by a single operator. At the same time, it obtained extensive information about nonsubscribers in those three systems—their demographic characteristics, their attitudes toward television and toward cable, their knowledge of cable, and their use of other related media. The project was conducted in the summer and fall of 1984.

Sites.   The three cable systems studied were small suburban cable systems in southern-central Michigan. The three systems had the following dates of initial service, numbers of subscribers, and penetration rates: System 1 began service in 1982, had 850 subscribers at the time the study commenced, a penetration rate of 62%; System 2 began service in 1983, had 350 subscribers, a penetration rate of 53%; System 3 began in 1983, had 680 subscribers, a penetration rate of 51%. All three systems had a 35-channel capacity with 23 channels then programmed, including three premium channels.

Research plan.   The plan developed targeted all nonsubscribers in the three systems as potential receivers of the free trial subscription. Working with street maps of the three communities and a complete address listing of all subscribers, an alphabetical list of streets in each coverage area was created, and the house numbers of all subscribers were listed. These house numbers were avoided by interviewers making the initial search and contacts with nonsubscribers only. The project then proceeded in two phases.

In the first phase of this study, in-person interviews, lasting approximately 20 to 25 minutes, were conducted with nonsubscribers by trained student interviewers from Michigan State University. The offer of free cable installation and 1 month of free cable service—all channels available in the specific system, including three premium channels—was made at the close of the interview. From this phase of the study, there emerged two groups of respondents for subsequent analysis—*Acceptors* and *Rejectors*. The former accepted the free cable offer, and the latter did not. Interviewers scheduled the installation of cable for the trial period for those who accepted.

The second phase of the study was a follow-up interview with those who had accepted the free trial offer. This interview was conducted by telephone and averaged 10 to 15 minutes to complete. It obtained additional information from each respondent, and also determined which cable service(s), if any, the respondent wished to retain. From this phase, two groups were created—*Keepers* and *Drops*—which will be discussed in the results section of this chapter. Interviewers then provided service/disconnect/downgrade information to the cable system for administrative handling.

Data from all three systems were combined to provide efficiency in reporting the results. There was a total of 1477 nonsubscribing addresses. Of these, 582 were personally interviewed for a 39% completion rate. The remainder consisted of unoccupied homes, no-contacts with an adult after three attempts, and refusals.

The final item in the first phase interview asked the respondents whether or not they would accept the free trial cable offer. Acceptors consisted of 369 of the

582 respondents, or 63% of those actually interviewed. Follow-up telephone interviews in the second phase of the study were conducted with all Acceptors, typically on the first day after the 30-day trial period ended.

Variables.   In order to ascertain the characteristics of nonsubscribers, and subsequently to differentiate Acceptors from Rejectors, and Keepers from Drops, among the nonsubscribers, the first phase questionnaire asked these nonsubscribers about the following:

Television in general.

- typical television viewing behavior, including hours per day viewed and number of sets available to them
- use of media other than television, e.g., newspapers, theatre, movies, and magazines.
- satisfaction with the variety and types of programming and services available in their community
- their characterization of television in general on a series of bipolar scales, e.g., boring–interesting, repetitive–varied, important-unimportant
- enjoyment of specific types of television program content
- their major sources of television program information, e.g., listings and guides.

Cable television.

- Prior cable television experiences, in particular previous subscribership and reasons for discontinuance.
- Exposure to cable in terms of watching at others' homes, receiving promotional literature, and having been contacted by a local cable company
- Reasons for not subscribing to the available cable service
- Suggestions as to how cable would be more attractive
- Recognition of existing cable channels
- Perceived importance of different components of cable, e.g., improved reception, more channels, movie channels.
- Interest in various cable channels and services available.

In addition, the Phase One questionnaire asked questions about the demographic characteristics of the nonsubscribers. Included were length of residence in the area, marital status, household size, whether the household had children, whether the residence was owned or rented, education, age, occupation, place of work, working hours of the principal wage earners, household income, and the gender of the respondent.

The Phase Two questionnaire administered by telephone at the end of the free trial period, probed into the Acceptors' experiences with cable in an attempt to discern why some of them chose to subscribe to cable now and others did not.

Many questions were identical or similar to those asked in Phase One, but they were now in the context of having had experience with cable for 30 days. In most cases, the same respondent interviewed at the start of the study was re-interviewed. Phase Two question areas for these Acceptors included:

- their cable television viewing behavior in the last month, e.g., hours per day, comparison with precable
- their characterization of the attributes of cable television
- the appeal (or lack of it) of cable television to them
- major sources of cable television programming information
- awareness of available cable channels, viewing of those channels, and favorite channels
- satisfaction with the three premium channels (HBO, Cinemax, and Disney) received gratis during the trial period
- the importance of various components of cable television to them, e.g., 24-hour programming, nonlocal stations
- suggestions for improving cable television

At the close of this interview, respondents were asked whether they wished to retain cable and be billed for it or not. Those who dropped cable were asked why, and those who continued it were asked which services they wished to retain. For the latter, monthly charges were identified for the services requested.

RESULTS

To review, there were four groups of respondents of primary interest in this study. First, those who accepted the trial offer at the conclusion of the first interview: *Acceptors*. Second were those who rejected that same offer: *Rejectors*. Results from Acceptors and Rejectors convey responses from the first interview. Third, those who accepted were re-interviewed as soon as possible after the 30-day free trial offer concluded, and invited to subscribe to cable. This yielded a group who accepted cable on a billing basis: *Keepers*; and a group who continued to abstain from cable subscribership: *Drops*. Results, where reported from Keepers and Drops, come from the second interview.

Subscription to cable.   Inasmuch as the primary purpose of the study was to assess the effectiveness of the marketing technique of a free 30-day trial offer of all cable services, the key findings are to be contained in the extent to which cable was subscribed to by those who were Acceptors.

In System 1, 64% of the Acceptors decided to keep cable; in System 2, 85%; and in System 3, 70%. Combining these three systems, with their different numbers of respondents (369 Acceptors altogether), the overall subscriber rate among these former nonsubscribers was 70%. This appears to be an impressive level of acceptance, and even more impressive when one recognizes that no other marketing efforts were implemented during this experimental period, e.g., no direct mail and no personal contact.

When placed in the broader context of all nonsubscribing households in the systems, including those not contacted in Phase 1, this experimental effort added 17% of the population of nonsubscribers to the subscriber rosters of the three systems combined. And in terms of the overall system penetration, there was a net increase of 8% of homes passed. Given that most marketing efforts which generate a few percent increase in subscribers (from among current nonsubscribers) would be considered successful, the present effort is that much more noteworthy. Table 1 summarizes the impact of the trial offer on system penetration.

To which services did Keepers subscribe? The options available included basic-only service, full service (extended basic), and any or all of three premium channels: HBO, Cinemax, and the Disney channel. The top choice was full service plus a single premium channel (31%), followed by full service only (25%), full service plus two premium channels (20%), basic only (11%), full service plus all three premium channels (9%), and basic with HBO (5%). Thus, the marketing experiment fulfilled a second goal of most cable system operators, generating premium channel customers in large proportions. Only one-third of Keepers did not subscribe to at least one premium channel.

Non-subscribers. Before looking closely at the four subgroups established, recall that all respondents in Phase 1 were nonsubscribers. At that time, they were asked about their reasons for not subscribing, why someone might be taking cable, and what would make them more likely to subscribe. Answers to these questions provide a suitable context for the remainder of this results section.

The dominant reason for nonsubscription was the belief that cable was too expensive, as stated by 32% of the respondents. Two other major reasons offered were that they did not watch enough television to warrant adding cable (15%) and they just had not "gotten around to it yet" (12%). No other single reason was offered by as much as 10% of the respondents.

Most prominent among reasons why someone might choose to subscribe was the greater variety perceived to be available on cable, reported by 25% of the respondents. Two other reasons seemingly related to the first response were also

Table 1. Subscriber/Penetration Level Activity During the Study

	N	%
Homes passed	3357	
Subscribers (penetration) @ Phase I	1880	56%
Non-Subscribers @ Phase I	1477	44%
Interviewed @ Phase I	582	
"Acceptors" @ Phase I	369	
"Keepers" @ Phase II	258	
Subscribers (penetration) @ Phase II	2138	64%
Increase: (penetration) Onset to Phase II		8%

given—greater entertainment (13%) and more movies (12%). Insignificant portions mentioned reception quality, fewer commercials, or 24-hour programming as reason for subscribing to cable.

When asked what would make them more likely to subscribe to cable, one-third replied that lower rates would do it, whereas one-fourth said that "nothing" would make it more likely. Again, no other response came from as much as 10% of the respondents.

Free trial offer groups.   The purpose of comparing Acceptors to Rejectors and Keepers to Drops is to provide information about them that may be useful in systems other than the ones studied here. Although this population of nonsubscribers may in some ways be idiosyncratic to the three systems whose data have been merged here, i.e., small towns in the midwest, when we are dealing with these alternative responses to the free trial offer, it is likely that we have more sensitive and useful information about segments of the nonsubscriber market. This is first exemplified in the demographic comparisons among these groups, as shown in Table 2. This table (and those which follow) has been constructed so that the data from Acceptors and Rejectors contain their responses at Phase 1, and the data from Keepers and Drops are from Phase 2, after having had cable for 30 days.

Those who accepted the trial offer, as compared to those who rejected the offer, were considerably younger and had therefore lived in their then-current residence a substantially shorter period of time. They had larger households, with more children in those households. Although a majority of Acceptors did own their home, it was a much smaller majority than among Rejectors of the offer. The Acceptors were also more likely to be high-school graduates, whereas

Table 2.   Demographic Characteristics by Subgroup

Characteristic	Acceptors	Keepers	Drops	Rejectors
Age	36.9	35.3	38.2	53.4
Length of Residence (years)	13.4	12.2	13.5	39.0
Household Size	2.99	3.14	2.90	2.38
Children in HH % Yes	46.2	56.0	48.0	17.0
Housing Status				
% Own	56.9	52.4	62.0	80.4
% Lease or Rent	43.1	47.6	38.0	16.2
Education Level				
% Less than HS	17.4	16.7	19.1	19.5
% HS Graduate	48.8	50.0	45.6	36.2
% Some College	22.3	21.9	21.1	27.0
% Some Grad	10.2	9.8	14.3	17.2
Annual HH Income				
% < $15,000	43.9	41.9	47.3	46.9
% $15K–$25K	28.7	30.4	26.5	17.1
% $25K–$35K	13.3	13.9	9.7	20.0
% $35,000 +	14.1	13.9	16.5	16.0

most of the Rejectors had some college education. Finally, the income level of the Acceptors was lower than those of Rejectors.

After the free trial cable period ended, Acceptors were subdivided into Keepers and Drops. The former were more likely to have children in their slightly larger, rented households. Education and income differences between these two groups were negligible. Thus, the key demographic differences were between those who did and did not accept the original free trial offer.

Television behavior.   Let us look first at how television was used, before and after the cable period. Table 3 provides a summary. At Phase 1, the least amount of television viewing per day was reported by those who rejected the cable trial, although they still reported an average of 3.5 hours viewed per day. Acceptors were already averaging 1 hour more per day than that level reported by Rejectors. However, the subgroup of eventual Drops were substantially lighter TV viewers than those who later chose to subscribe. At Phase 2, this difference in viewing levels persisted between Keepers and Drops, *both of whom showed large increases in their viewing between the first and second interviews.* This difference is recognized by them in their responses to the question of whether they had watched more, less, or the same amount since receiving cable in their homes; 64% of the Keepers reported more viewing, compared to 40% of the Drops. The subgroups did not differ in their access to television; all reported having equivalent numbers of television sets in their households. However, those who rejected the offer were more likely to watch television alone, a likely reflection of their age and their smaller households.

Television content.   Respondents indicated how much they enjoyed each of more than 30 different kinds of television content, e.g., news, movies, and soaps, on a scale ranging from 0 (no enjoyment at all) to 10. Across everyone, the highest ratings were given to old movies, new movies, state and local news, weather, and action-adventure programs, all averaging more than 6.0 on the enjoyment scale. Minimally enjoyed were financial news, and shows featuring jazz or classical music, all averaging less than 4.0. Those who rejected the cable offer

Table 3.   Television Viewing Behavior/Environment

Subgroup	Phase I	Phase II
	Average Hours of TV Viewing on a Typical Weekday	
Acceptors	4.61	
Keepers	4.83	5.57
Drops	4.17	4.98
Rejectors	3.59	

	Reported TV Viewing Since Cable		
	% More	% Less	% Same
Keepers	63.5	2.4	34.3
Drops	40.3	3.0	56.7

were more likely to give substantially *higher* enjoyment ratings to what might be termed informational programming—national, state and local news, weather, documentaries, religion, shows about government, science, and even sports programs. The Rejectors were much *less likely* to enjoy new movies, soaps, situation comedies, rock music, dance shows, cartoons, and re-runs of old series. Keepers and Drops were much more alike in their enjoyment ratings with few exceptions. So initial content preferences varied markedly between those who were willing to try cable on a free-trial basis and those who would not even do that.

One further attitudinal measure of television and cable used in this study was more generic than specific; it asked for judgments about television in general on a set of adjective-opposite attributes, e.g., good–bad, and asked for the same judgments about cable TV from those households who were cabled free for 30 days. The key results are in Table 4. For most attributes, among the four sets of data, little difference was reported at the time of the first interview. Rejectors considered television less good and more commonplace than Acceptors, but, of primary interest are the differences in the assessments made by the Keepers and Drops between the two interview phases. After a month with cable, the Keepers rated television as considerably more interesting, better, more gentle, more varied, closer to unique, more influential, and somewhat more relaxing and more pleasant than did Drops. The Drops were not unaffected during this period with cable; their ratings followed the same pattern as Keepers, but typically with smaller changes, except for their perception of the influence of cable. Thus, television developed a new and substantially more positive image among those who experimented with cable and especially among those who made the decision to retain it than among Rejectors and Drops.

Table 4.   Characterization of Television/Cable Television

Attributes	Acceptors	Keepers	Drops	Rejectors
Boring to Interesting	3.34	3.54	3.32	3.25
(Phase 2)		4.00	3.45	
Exciting to Relaxing	3.63	3.73	3.85	3.69
(Phase 2)		3.50	3.42	
Unpleasant to Pleasant	3.53	3.68	3.59	3.45
(Phase 2)		3.98	3.59	
Bad to Good	3.42	3.68	3.49	3.18
(Phase 2)		4.07	3.59	
Violent to Gentle	2.58	2.67	2.86	2.49
(Phase 2)		3.11	2.77	
Repetitive to Varied	2.43	2.52	2.58	2.60
(Phase 2)		3.19	2.89	
Commonplace to Unique	2.50	2.49	2.90	2.22
(Phase 2)		3.14	2.99	
Not Influential to Influential	3.62	3.68	3.71	3.41
(Phase 2)		3.20	2.91	

Cable channels.   A major portion of the questionnaire dealt with what was available on cable, what was of interest to the potential subscriber, and what features might be attractive. Beginning with what channels might be of interest, respondents were given a listing of 11 basic cable channels and six premium channels, and were asked how interested they were in each of them. For the basic channels, the specific channels were not named, but a generic explanation of each channel's programming content was provided, e.g., continuously updated national and international news, 24 hours a day (CNN). The premium channels asked about were not presently available in any of the systems studied. The pattern of results was extremely consistent. For each of the 11 basic channels described, Acceptors rated their interest higher than did Rejectors. The average difference in interest, across all 11, was 15% more Acceptors at least somewhat interested in the channels. The greatest interest difference between Acceptors and Rejectors was reported for superstations (WGN, WOR, and WTBS), where the difference was 26% to 32%, and minimal with the news channels (CNN, C-SPAN) where the difference was 4%. This of course was before any of the respondents had cable. Then, after the trial period, comparisons between Keepers and Drops showed a similar pattern, with the former reporting more interest in each of these channels, although the differences between Keepers and Drops were narrower, averaging 10% for each channel.

Among the half-dozen pay channels, a similar pattern was found, but it was more intense. On average, 23% more of those who accepted the trial offer were interested in what the channels had to offer than those who rejected the offer. And at Phase 2, the Keepers were more likely to be interested in each channel than those who dropped cable.

Recognition of cable channels was another aspect of the study. Here, respondents were simply asked if they had ever heard of 12 different channels, including three premium channels, all currently available in the systems studied. The channels most recognized by all subgroups were the premium channels. However, Acceptors were more likely to indicate that they had heard of every channel than were Rejectors. In both interviews, Keepers' and Drops' recognition levels were similar; however, there was a sharp increase in overall recognition by both groups after having cable for a month, particularly of those channels which were lesser known at the first interview, e.g., recognition of Lifetime increased from 17% to 45%, WTBS from 47% to 78%, and USA from 61% to 80%, between Phases 1 and 2.

What was actually watched, and what was enjoyed? Phase-2 respondents were asked if they recalled watching each of the channels available on their local system, a comparison between Keepers and Drops, given that Rejectors had no opportunity to view the channels. For many channels, the differences between these two groups were quite small, although a larger proportion of Keepers than Drops reported watching 18 of the 25 channels listed. The most viewed channels were the three premium channels; 80% viewed HBO and Cinemax, and 57% the

Disney Channel. Next came WTBS (44%), and MTV (44% of the Keepers and 30% of the Drops). The only other channel recalled by more than 30% of each group was ESPN. One consistent pattern of differences between Keepers and Drops was in their recall of viewing local off-air stations carried on the cable system, with an average of 10% more of the Keepers than Drops remembering having viewed such channels.

As for satisfaction with cable, sharper differences emerged, as reported in Table 5. First, the overall level of satisfaction reported was very high with each of the premium channels, with the basic channels generally, and with the service provided by the cable system. Across all these areas, the level of being at least somewhat satisfied ranged from 81% to 98%. However, the proportion of Drops not satisfied with each item is higher than for Keepers, and this dissatisfaction was found primarily for two of the premium channels. Although satisfaction with cable does not appear to be a general basis for Drops not continuing with the service offered to them, it does seem to impact this decision to some extent.

Cable features.   Which aspects of the cable offerings were considered to be more or less important for the households studied? What interest was expressed in extra or optional cable services that the system might subsequently develop? Table 6 gives the importance ratings from Acceptors and Rejectors at the time of the first interview, and from Keepers and Drops at the second interview. Essentially, Acceptors judge each of these features to be nearly equivalent in importance—improved reception, the large number of channels, 24-hour-a-day service, and the availability of movie channels, news channels, and independent stations—with at least 80% considering them at least somewhat important. Each of these features seems far less important to Rejectors of the trial offer, seeing that nearly 30% fewer of them indicated each feature to be important to them.

Table 5.   Satisfaction with Cable

Channel(s)/Service	Keepers	Drops
HBO		
% Satisfied	93.3	80.5
% Not Satisfied	6.7	19.5
Cinemax		
% Satisfied	89.1	80.7
% Not Satisfied	10.9	19.3
The Disney Channel		
% Satisfied	95.4	93.1
% Not Satisfied	4.6	6.9
Basic Channels		
% Satisfied	98.2	94.6
% Not Satisfied	1.8	5.4
Cable System Service		
% Satisfied	98.0	97.2
% Not Satisfied	2.0	2.8

Although those who chose to drop the cable service indicated consistently lower importance of features than did the Keepers, those differences were much smaller than between Acceptors and Rejectors.

In addition, respondents were asked about their level of interest in nine different services that conceivably could become part of their local cable system—a local system cable guide, home security/alarm systems, additional FM stations, home shopping and banking via cable, discounts on retail merchandise advertised on cable, closed circuit special events, a video games network, and a remote control channel selector. More than one-half of the Acceptors were interested in every one of these potential services, headed by their interest in a local cable guide (95%) and lagged by their interest in home banking (51%). Among Rejectors, interest in a cable guide was reported by 67% and interest in home security by 53%. For each of these services, Keepers were far more interested than the Drops, and Keepers were substantially more interested in these offerings than they had been before their experimental period with cable.

Cable fees.   Respondents were asked to estimate the installation fee and the average monthly fee for having cable, as well as the number of channels they would receive. Acceptors and Rejectors made similar estimates of installation costs; the former estimated the monthly fee as $1.60 *higher*; both made similar estimates of the number of channels available. Although cost was a key factor among Rejectors, it was not based on exaggerated estimates, nor estimates that varied greatly from those made by Acceptors of the trial period.

Television guides.   Before cable existed in these systems, all groups relied primarily on the newspaper and secondarily on *TV Guide* for television program schedule information. Reliance on both of these, but particularly the latter, was diffused by the availability of a cable system listing, which became the primary source of cable TV information for one-fourth of the Acceptors. Thus, there was a rapid shift in reliance on available sources of information by a quite substantial portion of the cabled households.

Other media use.   Respondents received an average of one newspaper and two magazines delivered regularly to their homes. They also attended an average of one movie in a theatre each month. Did this vary among our subgroups and was it affected during the trial cable session? Neither newspaper nor magazine

Table 6.   Importance Rating of Cable Features

	% At Least Somewhat Important			
Feature	Acceptors	Keepers	Drops	Rejectors
Improved television reception	91.1	95.1	84.2	72.7
Entertainment & Info. 24-hours/day	84.7	85.6	72.0	58.4
30 or more channels of TV	85.8	84.9	72.7	45.7
Movie channels, such as HBO	84.6	85.1	74.2	53.9
Independent stations, such as WTBS	79.5	80.8	71.8	42.4
News & Info. channels	86.8	80.9	78.0	74.5

reception was affected. Although Rejectors of cable were heavier magazine users and lighter moviegoers than Acceptors, both Keepers and Drops averaged less movie-going after their first month with cable than they had reported in the first interview; the Keepers averaged one less movie in that month, and Drops one-half fewer movies. Whether this would continue is indeterminate from this study, but it appears that at least the initial availability of movie channels substantially impacted on theatre movie attendance.

DISCUSSION

At the end of a 30-day free trial period with cable television in three systems, subscriber penetration had increased 7%, 8%, and 9%, reflecting acceptance by two-thirds of those who had accepted the initial offer. This was in systems whose penetration levels had been relatively stable for 1 to 2 years, from the time the system became operative to the time this study began. Two to 3 months after the free trial period ended, penetration levels were 5%, 7%, and 6% above those at the beginning of the study. This was accomplished under the following limitations:

- no promotional or sales contact by the cable firm was made with the respondents during the period of the study, save the distribution of a channel listings guide to those who accepted the trial offer, and, at the end of the Phase 2 interview, asking if they wished to retain any of the cable services obtained during the trial period.
- this portion of the final contact (as well as the initial contact) was incidental to the interview itself, consisting solely of an offer to be accepted or rejected.
- all respondent contacts were made by university students working as trained interviewers, not by professional salespersons.
- those who rejected the initial offer were not re-contacted during the study period.

The result: substantial success from a very low keyed sales presentation. Had this trial offer been combined with normal reinforcing strategies, e.g., follow-up advertising, direct sales, second attempts with those who participated in the trial but then dropped the service, or second offers to the initial Rejectors, the results might have been stronger, but then the specific impact of the free trial offer would have been confounded by other factors, which is why they were omitted.

On the other hand, perhaps it was because of this low-key, minimal sales, maximum information approach that these groupings of recalcitrant, reluctant nonsubscribers became more receptive to having cable in their homes. After all, cable had been available to them for 1 to 2 years, half or more of their neighbors had it, virtually all of them had seen what it looked like, and yet they had chosen not to subscribe, "had not gotten around to it," didn't think it was worth the money, etc. Yet, after a month of experimentation with cable, more than half of those nonsubscribers contacted chose to adopt it. This suggests another experiment, directly pitting the present strategy against that of hard sales, and comparing the cost per new subscriber between the two approaches, or some combination of them.

Perhaps, however, the Acceptors and subsequently the Keepers represent another nonsubscriber type that has not previously emerged in the cable TV research literature, one that operates on the principle of "I'm from Missouri (or Michigan), show me!" This would be akin to the trial demonstration often utilized in the diffusion of innovations, but on a much more personal basis. For example, in attempting to get farmers to adopt a new farming technique, area farmers might be invited to a single farm site where that technique has been utilized so that they can see the results obtained. That demonstration is one step removed from the strategy implemented here, where each could have the opportunity to experiment in his or her own home, at a much lower risk than the farmers'.

An important caveat is that Keepers in this study were only interviewed at the time of their initial decision to subscribe. Churn rates for these cable converts should be closely followed, to determine how their churn levels compare to overall system churn.

In this study, the first interview undoubtedly helped establish a good deal of rapport between interviewer and respondent and that may have increased the likelihood of acceptance of the trial offer. However, without that interpersonal contact, there would likely still be strong acceptance of such an offer—something for nothing. This should be the case among those who appear to be most susceptible—young, marrieds, with young children, with marginal incomes. Perhaps they need to see cable to believe it; perhaps they need to get it for free before they can decide it is worth the price; perhaps they need it around as part of the furniture before accepting it; perhaps it is a felt debt because of the gift (free trial) received. Whatever one's rationale for accepting and subscribing to cable, this free trial offer was a key inducement among an important and large segment of the cable market.

From those who accepted the original offer, but then dropped the cable opportunity when they had to pay for it, there is little in the results of this study to help with marketing cable to them. Drops are similar to Keepers in demography, content preferences, and attitudinal reactions to cable offerings. They were, however, somewhat less satisfied with what cable brought to them. Additional research is needed if one wishes to tap what would bring these Drops greater satisfaction; perhaps it is some cable channel or channels not currently available on these systems; perhaps for them a longer trial period (60 days) would be sufficient to move them into the Keeper ranks.

What of Rejectors, those who would not even have cable around for free? They are a conspicuously older group, and one for whom cable is affordable.

They submitted to an interview with a young college student which suggests their rejection cannot be entirely a function of fear of strangers knocking at their door. Perhaps it is some anxiety as to who will be knocking next—to 'force' them to buy the cable. Again, diffusion of innovation theory would suggest that using an older interviewer (or salesperson) more similar in appearance to the respondent might offer further credence in this situation.

Finally, there is the bulk of nonsubscribers—two-thirds in each system—for whom three contacts were insufficient to reach or who refused to be interviewed. Doing the field work in the summer led to many empty households (vacationing residents); for them, an alternative season might have been more appropriate to make a trial offer. For the rest, as well as for the Rejectors and the Drops, a variety of marketing strategies based on data presented in this chapter can be devised and attempted. One might focus on increasing awareness of cable's offerings. This could alter obvious areas of ignorance or misinformation among those who find fault with the content. A second approach might focus on the value received and be targeted to those who make primarily price objections. A third strategy might be to turn to those who have just "not gotten around to it, yet," and let subscription be at the initiative of the cable system rather than that of the potential customer. The data set would appear to provide many options to be tailored to individual system needs and the profile of nonsubscribers identified in each of those systems.

Whether what worked in these three systems will work elsewhere cannot be determined, for reasons already identified. Lacking arguments to the contrary, however, the best guess is that having the opportunity to try something out in one's own home, at one's own pace, with no extrinsic costs involved, is a useful and successful approach to marketing a quality product or service, such as cable television.

## CHAPTER 20

Free System-Specific Cable Guides as an Incentive

Bradley S. Greenberg, Roger Srigley, Thomas F. Baldwin, and Carrie Heeter

In the spring of 1984, Continental Cablevision of Lansing, Michigan, decided to explore the issues involved in distributing a system-specific cable television viewing guide. Such a guide would contain schedule and programming information for each channel in the local system; such guides were and still are infrequent among the several thousand cable systems in the United States.

In collaboration with Michigan State University, a study was designed to assess the impact of such a guide. In general, two major questions were of most import:

1. Would the presence of such a guide in the homes of system subscribers reduce disconnects and downgrades?
2. Would the content and design of the guide be favorably received by system subscribers; e.g., would it be used instead of alternative information sources, would it convey new information, would it increase the use of certain channels for which information was now more readily available, would it improve attitudes toward the cable company, etc.?

As a result of the joint interest of Continental and mass communication researchers at MSU, a field experiment was planned and conducted over a 5-month period. Beginning in August 1984 and continuing through December 1984, five issues of a system-specific cable television guide were produced and distributed to test groups of subscribers in the franchise area. Concurrently, three waves of interviewing were conducted by telephone in that market.

Prior to the guide's publication, six focus group sessions were conducted, involving 25 current cable subscribers, drawn from basic, one-pay, and two-pay categories. In these sessions, we examined their use of currently available guides in the newspapers, weekly guides they purchased, and monthly guides received by pay subscribers. Discussed was the comprehensiveness of information, ease of access to desired information, its accuracy, the format in which the information was presented, and the overall size and style of guides. Examples of different kinds of guides, in different sizes and styles, were presented, including some hypothetically designed by the research team.

Information obtained during the focus group sessions was evaluated by both Continental and the researchers. Suggestions considered critical were included in the design of the guide, e.g., the size of the guide, the inclusion of a grid identifying each channel, the creation of a primetime grid for all major channels, daily program listings for all possible channels, and the emphasis on the local origin of the guide.

Field surveys. The first field survey was conducted during the month prior to publishing the first issue of the new cable television guide. The second survey was completed after the third issue of the guide had been distributed, and the final survey was run after the fifth and final issue of the guide had been distributed.

The design of the three-wave experimental study was as follows:

1. For the first wave, a simple random sample of basic cable subscribers, HBO subscribers, and two-pay (HBO and Cinemax) subscribers was drawn from computer listings at Continental; this was supplemented by a telephone directory sampling of the Lansing area to obtain nonsubscribers. Interviews were completed with 209 basic, 104 one-pay, 109 two-pay, and 194 nonsubscribers. These represented completion rates of 72% among each of the three cable subscriber groups and 55% among the nonsubscribers.
2. The distribution of the test guides was made to a separate series of samples of the three types of cable subscribers. Computer-drawn random samples of 300 basic, 300 one-pay, and 300 two-pay subscribers were created, excluding those sampled in the first wave. These cable subscribers received five published issues of the test guide.
3. In Wave 2, interviews were attempted with each of those individual homes receiving the test guide. The completion rate was 72%; in all, 650 interviews were completed.
4. In Wave 3, attempts were made to re-interview the 650 respondents interviewed in the second phase. Interviewers attempted to re-contact the same individual in the household; otherwise, they interviewed a second eligible respondent in the same household. In all, 477 interviews were completed, a 73% completion rate.

Variables.   Two sets of data were important to this project: system data and questionnaire data.

To examine the impact of the guide in the test homes on system churn and downgrading, the homes receiving the guides were specially tracked, and all changes in the services received during the 5 months the guide was distributed were noted: Disconnects and downgrading rates were computed separately for the three subscriber groups receiving the guides. These rates were compared with those for the entire system of more than 47,000 subscribers during the same period of time, excluding the 900 homes forming the experimental groups.

A questionnaire was prepared and pretested by the research team in coordination with Continental. Interviewing was done by trained MSU students, working concurrently at a bank of telephones in a common room supervised by an MSU doctoral candidate.

The 20- to 25-minute questionnaire contained the following elements, included in all three waves, unless otherwise indicated:

1. Current receipt and use of specific, locally available daily newspaper television guides and magazine television guides, and an evaluation of those guides.
2. Awareness, knowledge about, and use of the Continental guide, including an evaluation of its content and format characteristics and the utility of specific sections of information (asked only in the second and third waves).
3. General use of television guides, in terms of when a guide is likely to be used and what kind of information the viewer is typically seeking; e.g., what the story is about, the length of the show, whether it's a repeat or not, etc.
4. Knowledge about the local cable system. A set of 16 questions constituted an index of respondent knowledge about the local cable system. Questions included costs of basic and pay services, the availability of R-rated and X-rated movies, radio stations, a children's programs channel, etc. (used in all three waves; however, in Wave 2, used with half the respondents, and used with the other half in Wave 3).
5. A measure of attitudes toward Continental Cablevision. Eight questions tapped satisfaction with a variety of issues related to the services received. For example, respondents were asked if it was hard to get hold of the company by phone, if the bills were confusing, if the company were interested in the local community, etc. (asked only of cable subscribers and split-halved in Waves 2 and 3).
6. Familiarity with and viewership of each channel on the cable system was assessed. For each channel on the system, respondents were asked how many days in the last 7 they had watched something. This was designed to provide a measure of relative use of different types of channels within the cable system. (In Wave 1, nonsubscribers were asked only about off-air channels; these questions were split-halved among respondents in Waves 2 and 3.)
7. Demographic questions were asked of all respondents in each wave. These included family structure and size, gender, ethnicity, income, and education, among others.

Let us turn now to major findings from this study.

THE SYSTEM AND ITS SUBSCRIBERS

The Guide and System Churn

Each month that the special cable television guide was published, it was distributed to the subsamples of households that had been randomly selected from computer lists of Continental Cablevision subscribers. In all, 900 households received the guide, 300 each from random samplings of basic, HBO-only, and two-pay subscribers.

Guides were published beginning in August 1984, and continuing through December 1984. Each month, the 900 receiving households were examined for churn—disconnecting cable completely or reducing the number of pay services received. We did not examine the incidence of upgrading which may have occurred. Churn from the households receiving the guide was compared with churn in the rest of the cable system.

In order to make the comparisons, it was necessary to weight the data from the experimental households so that it reflected the proportional distribution of subscribers in the entire system. For example, in August 1984, basic subscribers comprised 51.4% of all subscribers; HBO-only subscribers were 35.2%, and two-pay subscribers were 13.4%, of all subscribers. Churn in each group receiving the guide then was weighted by these proportions. In subsequent months, the proportions then found in the system were used. Here is how it worked in August 1984:

To examine *churn among basic subscribers*, the number of disconnects in each subscriber group was identified; there were six among the 300 basic-only subscribers receiving the guide, seven among the 300 HBO-only subscribers, and eight among the 300 two-pay subscribers receiving the guide. Thus, the formula used was:

$$6/300(.514) + 7/300(.352) + 8/300(.134) = 2.2\% \text{ disconnect}$$

To examine *HBO churn*, only the 600 HBO subscribers were relevant. Within the HBO-only guide recipients, there were seven complete disconnects (as included above) plus two downgrades to basic-only, or nine total downgrades. Within the two-pay guide recipients, there were eight disconnects (as included above), plus three downgrades to basic, plus one HBO disconnect, or a total of 12 HBO losses. The formula used was:

$$9/300 (.724) + 12/300(.276) = 3.3\% \text{ HBO churn}$$

The percentages used in this formula are the ratio of HBO-only subscribers to all pay subscribers.

To determine *total pay churn*, we looked at the total pay units lost among the 900 possible pay units in the 600 pay subscriber homes receiving the guide. In the example month being described, then, there were nine pay units lost in the HBO-only group and 26 pay units lost among the two-pay group (8 disconnects = 16 pay units; 3 downgrades to basic = 6 pay units; and 4 downgrades of one pay unit each = 4 pay units). For this measure, the formula used was:

$$9/300(.567) + 26/600(.443) = 3.6\% \text{ churn of pay units}$$

Here, the percentages used are the ratio of HBO-only units to all pay units. What then did we find?

Basic Churn

Month	System	Sample
August	1.4%	2.2%
September	1.6%	1.6%
October	2.9%	2.8%
November	2.1%	1.2%
December	1.2%	1.1%

Essentially, the pattern is inconsistent. In three of the months, but not sequentially, the disconnect rate in the system was virtually identical to that found in the sample receiving the guide. In one month, the system churn rate was larger than in the sample, and in one month, it was smaller.

HBO Churn

Month	System	Sample
August	2.4%	3.3%
September	2.5%	3.3%
October	4.3%	3.4%
November	3.5%	2.3%
December	2.3%	1.4%

Here, the pattern is consistent. For the last three months the guide was distributed, the HBO churn was smaller in the sample than in the system, averaging a 1% difference per month.

Pay Unit Churn

Month	System	Sample
August	2.7%	3.6%
September	2.8%	3.6%
October	4.9%	4.7%
November	3.9%	2.2%
December	2.5%	1.7%

Here, the pattern of results is similar to that found for HBO churn. During the first two months, the sample churn rate was larger than the system rate; by the third month, it was equivalent; for the final two months, downgrading in the sample was lower than in the system.

In all these comparisons, the month of August is an anomaly. Sample churn was larger than system churn. By October, the sample churn rate in all three groups was either lower or equivalent, and that pattern continued during November and December, most typically in the direction of a lower churn rate in the sample receiving the special television guide. If anything, this suggests that the impact of the guide was incremental, building slowly, especially among those subscribing to one or two pay services. The original plan for this study called for 6 months of experimental guide distribution, but the sixth month was deleted for budgetary reasons. With the proper hindsight now available, it seems that another 3 or 4 months of guide distribution would have been important in determining whether the trend which appears to have been identified in these results would have continued. Clearly, the October through December results are those which were sought in the conception of the system-specific cable television guide.

Demographics

This section differentiates among the subscriber groups and between the subscribers and nonsubscribers in terms of demographic characteristics. Of course, there is no expectation that the presence or absence of a guide was in any way related to these attributes. However, it does enable consideration of obtained demographic differences in planning marketing campaigns for different goals—to induce initial subscribership, to upgrade from basic to pay, and to upgrade one-pay subscribers to multipay. It also facilitates examination of group differences in succeeding sections of this chapter.

These are key differences (in Table 1) between nonsubscribers and subscribers:

1. The total household size among nonsubscribers is smaller, about 2.5 persons per household compared to an average of three in subscribing homes; nonsubscribers are less likely to have children, averaging less than one child per household.
2. They have fewer wage earners in the household; It follows, then, that the average income level in nonsubscribing households was lower.
3. The survey also showed that nonsubscribers had fewer working television sets in their homes (1.6) compared with subscribers (2.0).

Basic subscribers also differed from pay subscribers along a number of noteworthy attributes:

1. Basic subscribers had smaller households than pay subscribers, although the number of children was not consistently different. Basic households had less than three persons, and pay households three persons or more, on the average,

Table 1. Variable: Demographics

	Wave 1				Wave 2			Wave 3		
	Nonsubs	Basic	1-Pay	2-Pay	Basic	1-Pay	2-Pay	Basic	1-Pay	2-Pay
% Married	51	63	58	56	61%	62	61	64%	68	65
# of Children	.7*	1.7[ab]	1.1[a]	1.2[b]	.9	1.0	1.0	.9	1.1	1.2
Size of Household	2.4*	2.9	2.8	3.2	2.7[ab]	3.0[a]	3.1[b]	2.7[a]	3.0	3.3[a]
Age	42	42[ab]	38[a]	35[b]	45[ab]	38[a]	37[b]	46[ab]	39[a]	37[b]
Education-Self[1]	4.9[b]	4.6[ab]	4.8	5.0[a]	4.8	4.9	4.9	4.7[a]	5.0[a]	5.0
Education-Highest[1]	5.2	4.9	5.2	5.3	5.1	5.3	5.1	5.1[a]	5.3	5.5[a]
Full-time employed[2]	1.2*	1.5[ab]	1.9[a]	1.8[b]	1.2[ab]	1.5[a]	1.6[b]	1.1[ab]	1.5[a]	1.5[b]
Part-time employed	—	—	—	—	.4	.3	.4	.4	.4	.5
% Female	50%	60	61	56	50%	49	48	54%	47	54
% White	87%	87	88	82	89%	90	82	90%	91	81
Household income[3]	2.3*	2.5[ab]	3.0[a]	3.1[b]	2.8[ab]	3.2[a]	3.2[b]	2.7[ab]	3.4[a]	3.5[b]
# Working TV sets	1.6*	1.9[a]	2.2[a]	2.0	—	—	—	—	—	—
# Cabled sets	—	1.2[ab]	1.4[a]	1.4[b]	1.1[ab]	1.3[a]	1.4[b]	1.1[ab]	1.2[a]	1.4[b]

[1] 1 = <8th grade. . . .8 = Graduate degree
[2] At Wave 1, respoondents were asked how many wage earners there were
[3] 1 = 0–$14,000. . . .8 = >$75,000
*Nonsubs are significantly different from subs
[a,b]Means with the same superscript are significantly different from each other.

2. Educational level was lower in the basic subscriber households; in pay households, there is a year or two of additional college-level education. In fact, basic subscribers had lower educational levels than did the nonsubscribers.
3. Basic cable households had fewer persons employed full time; incomes were consistently lower in basic cable homes, compared to pay homes.
4. Basic subscribers are older than pay subscribers, averaging 43 or 44 years, compared to under-30 averages for pay subscribers.
5. In basic homes, fewer sets are connected to cable.

No consistent differences in terms of demographic characteristics are found between one-pay and multipay households. However, the data do suggest that minority households are more likely to be multipay subscribers; in communities where the proportion of minorities is greater, this difference may be larger and more meaningful. The only demographic characteristic included in this study which does not differentiate nonsubs from subs, and/or basic from pay subs, is marital status.

GUIDES

Guide Information Seeking

A series of questions asked respondents what kinds of information they sought when they were looking for information. These were not specific to the Continental Guide, but dealt with their more general searching needs. Table 2 summarizes the results.

Respondents were asked, "When you do check some TV listing, how often are you trying to find out what's on at a particular time?" The average responses are very close to the "often" response category. But there are systematic differences among the viewing groups: Nonsubscribers are least likely to seek this information, pay subscribers are most likely to seek this information, and basic subscribers are less likely to seek this information than pay subscribers.

This suggests that greater program selectivity is occurring as one moves from off-air only situations to having pay channels available from which to select. Although not statistically significant, the same trend emerges in terms of the respondent's use of a guide to find out about a particular show.

For the remainder of the items in Table 2, respondents were given this stem: "When you look for information about a show you might want to watch, what are you usually trying to find out . . . Are you trying to find out (e.g., what time the show starts)." Yes and No responses were requested for the nine issues asked about.

Although there are some between group differences, we find no consistent trends over time, and the information sought by nonsubscribers is similar in mag-

Table 2. Variable: Information Sought

	Wave 1				Wave 2			Wave 3		
	Nonsubs	Basic	1-Pay	2-Pay	Basic	1-Pay	2-Pay	Basic	1-Pay	2-Pay
What's on?[1]	3.4*	3.2[a]	3.2[b]	2.8[ab]	2.9[a]	2.5[a]	2.7	3.0	2.8	2.7
Find one show[1]	3.3	3.4	3.2	3.1	3.2	3.1	3.0	3.1	2.9	2.7
% who want to know:										
Time	92%	91	89	89	91%	94	86	89%	86	87
Name of Show	66	67	68	73	69	64	69	58	71	67
Channel	87	89[a]	82[ab]	94[b]	90	85	90	90	89	81
Length	58	52	57	56	65[a]	62[a]	46[ab]	53	44	56
Story Line	75	80	85	86	77	82	82	74[a]	76[a]	94[a]
Stars	42	44[a]	48	56[a]	44	36	42	41	37	43
Guests	40	44	41	42	44	38	48	37	32	46
If it's a repeat	52	54	60[a]	45[a]	57	60	51	56	63	46
Other shows on	84	77[ab]	88[a]	87[b]	80	85	83	79	81	85

[1] 1 = Very often; 2 = Quite often; 3 = Often; 4 = Not very often; 5 = Not at all
[a,b] Means with the same superscript are significantly different from each other.

nitude to that sought by cable subscribers. It is informative to identify the relative order in which these information needs are sought:

Eighty to ninety percent of the respondents are usually trying to find out—

> what time the show starts
> what channel it's on
> what the story is about
> what other shows are on

The story line appears to be of most interest to multipay subscribers, and what other shows are on to both sets of pay subscribers.

Next in terms of information seeking is the name of the show, consistently sought by about two-thirds of the respondents.

Slightly more than half the respondents want to determine the length of the show and whether it's a repeat or not. For both items, there is some tendency for the multipay subscribers to seek these pieces of information less often than any other viewing group.

Knowing who the stars or the guests on the show are is typically not sought by a majority of the viewers, averaging about 40% as an information need.

In designing a cable information guide, these clearly identify the information priorities of the viewers. The priorities appear to be quite stable over time. They did not vary as a function of the Continental guide being available, and indeed, they appear to have the same priority among those without cable as among those with cable.

Daily Newspaper Television Guide Use

To establish a context in which use of the Continental television guide could be assessed, we determined how newspaper and magazine television guides were being used.

Respondents were moved through a series of probes in assessing their use of television guides in their daily newspapers. First, they were asked for the names of the daily newspapers they received; second, they were asked which newspaper television guide they used most; and third, they were asked how often they used it, how easy or hard it was to use, how good or bad they found that listing, and if they kept it around to look at when they wanted to watch television.

First, the distribution of newspapers within the samples reveals these patterns:

1. Nonsubscribers to cable are least likely to subscribe to any newspaper. Fully 40% of the noncable households subscribed to no newspapers.
2. Nonsubscription to a daily newspaper is extensive among cable subscribing groups as well. On the average, one-third of cable subscribers do not take any newspaper on a daily basis.

3. The most immediate local newspaper dominated local subscription, a not surprising finding; it was subscribed to by from 50% to 66% of the sample groups in Wave 1 and 58% to 69% in Wave III; one paper from the nearest metropolitan area (Detroit) drew 10%–19% subscribership.

What of use of the newspaper television guide? After the distribution of the system-specific guides began, pay subscribers were less likely to use any newspaper television guide by comparison with basic subscribers; for example, in Wave 2, 36% of the pay subscribers did not use a newspaper television guide, compared with 22% of the basic subscribers.

The daily newspaper guide, regardless of which one it is, is used between once a day and four to five times a week by those who say they use one, as reported in Table 3. The multipay viewers are most variable in their response to newspaper guide use. During Wave 1, they were the highest users; in the Wave 2 sample, they were significantly less frequent in their use than any other cable sample group; and in Wave 3, their rate of use was equivalent to all other groups. This is further support for the notion that they move easily away from, and back to, the newspaper as a source of television information. Note also that they were least likely to keep the newspaper guide around to check on shows, but only at Wave 2.

In terms of ease of using the newspaper TV guides, the general response across all subscribing groups in all time periods was that they were "easy." This was an average score of 3 on a 4-point scale. Noncable subscribers tended to find them easier than any other group; this makes sense inasmuch as the newspaper guide is likely to have more comprehensive information about off-air channels, and it is more difficult for cable viewers to find cable information.

An overall judgment of the newspaper's television listing was elicited by asking the respondents how good or bad the listing was. There were no significant differences among viewing subgroups, and no differences across the time periods. The summative evaluation of these listings was that they were between "good" and "so-so," leaning toward "good," 2 on a 5-point scale. So, those who use newspaper listings as a major source of information—and many do not—also find them easy to use and evaluate them positively, all internally consistent behaviors and attitudes.

Magazine Television Guides

Parallel to the questioning done for newspaper television guides, the same series of question was asked for television guides in magazine form. These include magazines found in weekend newspapers, cable magazines distributed by pay cable channels, and *TV Guide*. Furthermore, by the second wave of this study, the experimental Continental guide was being distributed; a primary purpose of this series of questions was to determine the initial usage of the Continental guide

Table 3. Variable: Daily Newspaper Guide

	Wave 1				Wave 2			Wave 3		
	Nonsubs	Basic	1-Pay	2-Pay	Basic	1-Pay	2-Pay	Basic	1-Pay	2-Pay
Frequency of use[1]	2.34	2.71	2.27	2.13	2.59[a]	2.94	3.19[a]	2.81	3.07	2.78
Ease of finding information[2]	3.21	3.09	2.79	2.96	2.88	2.99	2.90	2.88	2.96	3.03
How good is listing[3]	2.27	2.28	2.15	2.61	2.22	2.24	2.31	2.33	2.32	2.32
% who keep it around	74%	81	80	87	72%[a]	61	53[a]	58%	60	67

[1] = Several times a day; 2 = once a day; 3 = 4–5 times/week; 4 = 2–3 times/week; 5 = less
[2] = Very hard. . . .4 = Very easy
[3] = Very good. . . .5 = Very bad
[a,b]Means with the same superscript are significantly different from each other.

in juxtaposition with the other available magazine guides. Comparisons of level of usage of magazines guides in contrast with newspaper guides will also be made.

Looking first at Wave 1, prior to the introduction of the system-specific Continental guide, there were significant differences in use of *TV Guide*, the local newspaper's television magazine, the pay cable magazine, and the extent to which no television magazine at all was used. The differences were as follows:

1. *TV Guide* was received by significantly more multipay households and by significantly fewer noncable households.
2. The local newspaper's TV magazine was received by significantly fewer noncable than cable households.
3. Pay cable magazines were received more by multipay than single pay.
4. Three-fifths of the nonsubscribers received no television magazine, compared to one-fifth of the multipay households, with the basic and one-pay households falling in between.

In Wave 2, the initial impetus of the Continental Guide was made, and was found differentially among the subscriber groups. The multipay group was significantly more likely to acknowledge receipt of the Continental guide, and at a level that paralleled use of *TV Guide* and the pay cable magazine.

For each magazine guide, the multipay group was more likely to recall receiving them and the basic subscribers least likely. And here, the basic group is significantly more likely to say they receive no guide. In fact, more of them say that than say they receive any single guide.

Next, we examined which television magazine guide was used most. In the first round of interviews, 60% of the nonsubs and basic subscribers were most dependent on *TV Guide*, compared to little more than 40% of the pay subscribers. Only one-sixth of the multipay subscribers relied on the local newspaper; in Wave 1, they were primarily dependent on the pay cable guide for their information.

By Wave 2, in just the second month of distribution of the Continental guide, one-tenth of the basic and one-pay subscribers said they were now relying primarily on that guide. And fully one-fifth of the two-pay subscribers used the Continental guide more than any other television guide. This came primarily at the expense of the pay cable magazine and *TV Guide*, on which they now relied much less.

And by Wave 3, the multipay subscribers relied on the Continental guide to the same extent as they did TV Guide, and only marginally less than hey relied on the pay cable magazine. For the one-pay subscribers, there was substantial attrition from the pay-cable magazine to the Continental guide. Among all pay subscribers, fully one-fourth of them now reported that the television magazine they used most was the Continental guide.

This strongly suggests an early and building impact of the Continental guide on cable subscribers, particularly pay subscribers. If the same trend had continued for another month or so—if the guide had been available—it seems likely that the Continental guide would have been primary among all other available television magazine guides for the cable subscribers in this system.

Table 4 reports on the extent to which these viewers used the various magazines, how easy it was for them to find information, an overall evaluative judgment, and whether or not they kept it around to check when they watched television; this was for the television magazine they used most. These findings are of importance:

1. Multipay subscribers are more frequent users of their television magazine guides, particularly more than basic subscribers.
2. At Wave 1, multipay subscribers reported significantly more difficulty in finding information in their television magazines than did basic subscribers; by Wave 2, this pattern was reversed, and the basic subscribers were reporting that it was harder to find information they wanted. The shifts in magazine dependence reported above may account for this change. However, among all viewers, the magazine guides were rated as "easy," leaning toward "very easy" in terms of finding what they wanted.
3. Multipay subscribers are most likely to keep their magazine television guide around to check when they watch television, and basic subscribers are least likely to do so. However, even among the former, 80% do keep it around to reuse.
4. Overall evaluation of the television magazine guide was strongest among the multipay subscribers, generally rating it as "quite good."

In making comparisons between use of newspaper television logs and magazine guides (between Tables 3 and 4), these differences emerge:

1. About one-third of the cable subscribers receive no newspaper and one-third receive no magazine guide. However, although the lack of a newspaper is uniform across cable subgroups, the lack of a reported television magazine is maximal among basic subscribers and minimal among two-pay households.
2. Although the local newspaper predominates among those who use a newspaper for television information, no single magazine is predominant. And the introduction of a new magazine, the Continental guide, swiftly made significant inroads in terms of becoming the magazine of primary use.
3. In all time periods, the magazine guide was more frequently used during the week than the newspaper guide.
4. In all time periods, the magazine guide was evaluated as easier to find information and as a better listing. By Waves 2 and 3, the magazine guide was more likely to be kept around to check.

Table 4. Variable: TV Magazine Use

	Wave 1				Wave 2			Wave 3		
	Nonsubs	Basic	1-Pay	2-Pay	Basic	1-Pay	2-Pay	Basic	1-Pay	2-Pay
Frequency of use[1]	2.43	2.99[a]	2.55	2.51[a]	2.40	2.46[a]	2.14[a]	2.45	2.47	2.37
Ease of finding information[2]	3.28	3.40[ab]	3.00	3.18[b]	3.09[ab]	3.28[a]	3.26[b]	3.24	3.18	3.16
How good is listing[3]	1.95	2.10	2.17	2.08	1.95[ab]	1.78[a]	1.80[b]	1.85	1.85	1.82
% who keep it around	83%	84	83	94	80%[ab]	91[a]	94[b]	81%	85	86

[1] = Several times a day; 2 = once a day; 3 = 4–5 times/week; 4 = 2–3 times/week; 5 = less
[2] 1 = Very hard. . . .4 = Very easy
[3] 1 = Very good. . . .5 = Very bad
[a,b]Means with the same superscript are significantly different from each other.

Continental Guide

The largest portion of this survey was created to tap responses to the system-specific television information guide. Knowledge, usage, attitudes, and utility were assessed during Waves 2 and 3 among guide recipients. By Wave 2, the cable subscribers in this experiment had received three issues; at Wave 3, they had received all five issues.

After three issues of the guide, it was acknowledged as being seen by 86% of the multipay subscribers and by two-thirds of the basic subscribers. By Wave 3, those differences were diminished and the guide had been seen by three-fourths of the basic subscribers. However, the same pattern emerges in terms of asking how many issues had been received. Multipay subscribers recall having received four or five issues, while basic subscribers recalled three or four issues.

Five informational items about the guide were asked. There were no significant differences across the subscriber groups. What we find is that: 90% knew that it came monthly and that it was free; two-thirds knew that the guide originated with Continental Cablevision; one-half were aware that it was a local guide; and one-third reported that it came at the end of the month during Wave 2, and this increased to 40% by Wave 3. A similar percentage indicated that it came at the beginning of the month; mail delays likely made this an accurate response for a portion of the sample.

A second series of questions dealt with attributes of the guide, as judged by the receivers:

1. In each wave, two-thirds of the subscribers wanted the guide to continue. In each case, the multipay subscribers were significantly more likely to want the guide to continue than the basic subscribers.
2. Eighty percent said the digest size of the guide was the ''right size'' for such a guide; two-thirds said they could get information about programs on all the channels they received.
3. The guide was judged to be accurate, easily readable, easy to locate information desired, and with an overall summative judgment of being a ''good'' guide. Each of these responses was one scaled response removed from the maximum positive favorable reaction.
4. About one-fourth indicated that they found out about new channels that they didn't know about.
5. In Wave 2, 10% said they found out about news programs they didn't know were available, and this increased to 14% in Wave 3.

The final section of questions about the guide focused on what kinds of information were sought and used.

1. The guide was used once or twice a week by each of the subscriber groups on the average.

2. The prime-time charts and the daily program listings were used most, again once or twice a week, with the latter used more so by the multipay subscribers.
3. Infrequently used were the channel charts on the back cover which provided the channel number, logo, and name of each channel. However, this particular feature of the guide was discontinued after the first two issues.
4. The pay movie information was used most frequently by the multipay subscribers in both waves, but matched by the one-pay subscribers in the third wave.

Finally, we asked how useful certain special sections of the guide were, including some that had been assessed in terms of frequency of use:

1. Overall, the guide was judged as "useful."
2. Special sections of the guide were consistently judged as more useful by the multipay subscribers than by the basic subscribers. These included the primetime program charts, the daily program listings, the channel chart identification on the back cover, and the pay movie information.
3. Generally considered not very useful were the crossword puzzle and the guide-to-the-guide.

In sum, there was a consistently positive evaluation about most components of the guide. Although it was not used on a daily basis, it was still considered far more useful than not. Correct attribution of the guide to Continental, or its localness, was not as strong as would be desired. But the more favorable response of the pay subscribers, particularly the multipay, to all aspects of the guide's content, structure, and format is worth identifying once again, and they used it more often for more different informational functions.

KNOWLEDGE AND ATTITUDES

Knowledge about Continental

A series of questions was created that tapped knowledge about the local system—knowledge about costs, offerings, special channels, the availability of radio on cable, etc. In all, 16 items were used in each wave of the survey, and the guide contained information in each issue about some of these topics. Generally, it was anticipated that knowledge would improve over time, as the guide was disseminated and read.

In Wave 1, the nonsubscribers knew significantly less (scores averaged 7.5 correct items of the 16) about Continental than did any of the cable subscriber groups (who averaged 9 correct). And, in turn, the basic subscribers knew less than did either of the pay cable groups. So, at Wave 1, the more you subscribed to, the more knowledgable you were about the system. This pattern persisted at Waves 2 and 3. In those survey periods, the basic subscribers knew significantly less than the one-pay subscribers, who in turn knew less than the two-pay households. At Wave 3, the basic subscribers could answer about 8 questions correctly, the one-pay averaged 9 correct, and the two-pay averaged scores of 10. There were no consistent differences over the three time periods, except among the multipay subscribers, whose scores improved modestly but persistently across time.

Pay cable subscribers were more knowledgeable in those content areas of central interest to them in all three time periods: Multipay cable subscribers were more accurate in identifying the two premium channels on the system, in identifying the correct range of the additional costs for those two channels, and in stating that you could get R-rated movies on this cable system.

All these are relative differences. Let us identify those knowledge areas of maximum, intermediate, and minimum accuracy:

1. Eighty percent or more of the cable subscribers knew that they could get radio stations on their cable system, that they could get R-rated films, that they could not get X-rated films, that there was a children's channel available, that local businesses could buy commercials, and that HBO was available.
2. Two-thirds of the cable subscribers knew that you could get Congress on one channel, and that there was no channel all about pets.
3. About one-half knew that Continental was part of a national company and not locally owned, knew that people in Lansing could make cable TV shows for free if they wanted to, knew the correct cost of basic service, and knew the correct cost of adding HBO to their basic service.
4. There was maximum confusion in terms of how many channels were available (only about 30% correctly said there were 30 to 35), whether you could get all the channels you would get if you didn't have cable, and whether Cinemax was an available pay channel and what it cost.

Overall, then, the information known by the cable subscribers was not particularly substantial. Cable subscribers were most likely to know those things with which they were most involved, e.g., pay services information. Knowledge was related to the extensiveness of their cable services—those without knew least, and those with the maximum services knew the most.

Attitude toward Continental Cablevision

During each wave, respondents were asked eight questions designed to tap the extent to which they were satisfied or dissatisfied with a variety of issues related to the service they were receiving:

They're hard to get hold of when I want to reach them.

The bills are confusing.

They are interested in the local community.

They keep changing the channels where programs are on the system.

They are interested in what I want to see.

It's too hard to find out what's on the channels.

It's easy to change or stop the service I get.

They didn't explain enough to me about how to find the different channels.

For each item, the response categories were "strongly agree, agree, disagree, and strongly disagree." Items were scored so that, the higher the score, the more positive the attitude. We also constructed an overall attitude index, by summing responses to the set of eight items.

The results indicate that these attitudes were largely unaffected during the experimental guide period. They were no different at Wave 3 than at Wave 2 or Wave 1. Furthermore, the attitudes of basic subscribers are the same as those of both one-pay and multipay subscribers in all three time periods.

The overall attitude index had a possible range of 8 (extremely negative) to 32 (extremely positive). The middle of this scale would then be 20. The range of attitudes found across all waves and all groups was 22.2 to 23.3. What this means is that the subscribers' overall attitudes toward the local franchise-holder were modestly positive.

For seven of the eight items, the average response was 3.0 on the 4.0 scale. Respondents consistently disagreed with such items as, "The bills are confusing," and equally consistently disagreed with such items as "It's easy to change or stop the service I get." For one item, there was greater ambivalence, where the average score indicated that half the respondents agreed and half disagreed: "They're hard to get hold of when I want to reach them." Thus, if one wished to target the single area of discontent, this would be it.

CHANNEL FAMILIARITY AND USAGE

Channel Familiarity

During each phase of the survey, respondents were asked:

"We would like to know which channels people are familiar with. Would you please name all those channels you remember being able to get?"

There was no aided recall; i.e., no channels were named for the respondents. The only prompting was to say, "Any others?" twice after they ceased naming channels.

An overall index of familiarity was created by summing the total number of

35 possible different channels named by each respondent. The results indicate that the multipay subscribers were familiar with three to five more channels than the basic subscribers, and tended to be familiar with more channels than the one-pay subscribers.

The basic subscribers were as familiar with the eight local off-air channels as the pay cable subscribers; the local independent station and the nonlocal network affiliates that were received off-air were less likely to be mentioned by the noncable households. Those latter stations required stronger antenna for adequate reception.

There is a consistent pattern of differences among the cable subscribers: The pay subscribers were more likely to be familiar with any channel than the basic subscribers. This pattern is found for four of the eight off-air channels and for a dozen of the cable-only channels, at least at Wave 2. By Wave 3, there was a general increase in channel familiarity among the basic and one-pay subscribers, so that familiarity was more approximately equivalent. One would like to attribute this enhanced familiarity to the information contained in the experimental guide.

Eighty percent or more of the pay subscribers spontaneously cited the pay channels. The local commercial network affiliates were cited by 70% or more of the subscribers. There was 50% familiarity for the nonlocal network affiliates, a local independent, a Detroit independent, and a Detroit PBS station, ESPN; 40% familiarity for a local PBS station and WTBS; and 30% familiarity for CNN, USA, and MTV. All other stations were spontaneously recalled by a smaller proportion of the subscribers.

Channel Repertoire

Interviewers read through the entire list of channels available on the cable system for those respondents with cable, and went through a list of off-air channels available to nonsubscribers. For each channel, the same question was asked: "In the last 7 days, how many days did you watch something on Channel __?" This information is not designed to indicate precise usage of the channels, but is designed to gauge the relative use of the different channels.

First, an overall index of how many different channels were viewed was constructed. Second an index of the total days of viewing across all channels identified by the respondents was created.

Without cable, an average of four different channels had been viewed sometime within the previous 7 days, from among the eight channels available. Basic cable households had watched 11 channels, one-pay had watched 12 channels, and two-pay had watched 13 channels in the 7 previous days from among the 35 available to them. It is apparent that the difference in pay cable households is merely the additional pay channel(s) subscribed to. There were no significant changes across the time period of the study.

The total days viewing measure (the sum of the number of days per week all channels are viewed) indicated that nonsubscribers watched significantly fewer days, but they were asked about far fewer channels. Within the cabled households, substantial differences persist. Basic households averaged 43 to 45 viewing days, one-pay averaged 47 to 52 viewing days, and multipay homes averaged 50 to 55 viewing days. In sum, there was more television viewing in the multipay households across more different channels.

Among the channels available to both the cabled and uncabled households, the general pattern is for the cabled households to report more viewing of those local and nearby channels than the uncabled homes. They watch the nonlocal network affiliates more frequently, as well as the local affiliates.

What channels receive the most viewership? Two stations were watched four or more times per week, on the average—the local CBS and NBC affiliates; HBO was watched three times a week by the one-pay households and four times a week by the two-pay households. Cinemax was watched nearly as often by those having it as was HBO.

The nearest ABC affiliate and a Detroit independent station were watched three times each week. Watched 2 to 2.5 times per week were the nonlocal CBS, NBC, and ABC affiliates, a local independent station, CNN and CNN-Headlines, the Atlanta superstation, MTV, and ESPN. Watched one to two times per week were: the local PBS station, a nonlocal PBS station, USA, the Nashville network, the weather channel, Lifetime, and Arts/Nickelodeon.

These are just about one-half the channels in the cable system. No other channel received average reported viewership of at least 1 day a week, although CBN could be rounded up to that level. Essentially, the rest of the channels or half the system were reported to be watched .1 or .2 days out of the previous 7.

One channel—MTV—showed a significant pattern of differences across the three viewing groups in each wave of data collection. Pay homes watched MTV more than basic homes.

Local network affiliates do not appear to suffer in these cabled homes. Those with cable do more overall viewing across more different channels but appear to retain their viewership of the network stations. If anything, there is an overall network gain because of increased viewing of nonlocal network affiliates. Only among the multipay households is there a tendency to decrease viewing of local network affiliates.

CONCLUSIONS

It may be useful to review some of the general objectives of a system-specific guide and determine if the study data indicate achievement of these objectives. These are:

1. To improve retention
    a. to reduce retention marketing costs
    b. to increase satisfaction
    c. to improve knowledge of cable
    d. to improve awareness of cable channels
    e. to expand use of channels
2. To encourage upgrades
3. To produce additional revenue
    a. through advertising sales
    b. through subscription sales

Retention.   The most obvious business value of the cable guide is to improve retention. And, the most direct measure of this effect is *churn*. Over the 5 months of the free guide trial, the presence of the guide did not seem to have a major influence on the churn of basic subscribers. However, among HBO-only subscribers, churn was down substantially during each of the last 3 months of trial. This is also true for the last 2 months among the two-pay subscribers.

It is reasonable to attribute this reduction in churn to the guide. The fact that the effect is noticed in the latter months of the trial could be expected. The guide's value to subscribers would register slowly as they became accustomed to its utility.

Why basic subs were not affected by the guides in the same magnitude as the pay subs is not clear. The guide had a much lesser overall impact on the basic subscribers. Perhaps because they are paying less, it is of less importance to them to get the most out of cable. Basic subscribers simply do not have the commitment of one- and two-pay subs. Clearly the pay subscribers are heavier TV users, and, because they are better educated, may be more selective. There is also the possibility that some of the basic subs upgraded to pay services as a result of increased awareness of that programming through the guides.

If we were to take only the last 3 months of the trial period, where the guide impact was beginning to be realized, we would see an improvement in revenues. Over those 3 months, the reduction in basic-only churn amounted to 280 subscribers, when the sample result is projected to the entire system. The revenue saved would be $2,730.[1] The reduction in HBO-only subs projected to the entire system was 456, a saving in revenue of $4,993.[2] The churn for two-pay subs was

---

[1] .001 (reduction in churn) $\times$ 25,185 subs (October) = 25
.009 (reduction in churn) $\times$ 25,574 subs (November) = 230
.001 (reduction in churn) $\times$ 25,098 subs (December) = 25
280 $\times$ $9.75 = $2,730 $\times$ 4 (quarters) = $10,920 savings per year
[2] .009 (reduction in churn) $\times$ 15,337 subs (October) = 138
.012 (reduction in churn) $\times$ 15,299 subs (November) = 184
.009 (reduction in churn) $\times$ 14,927 subs (December) = 134
456 $\times$ $10.95 = $4,993 $\times$ 4 (quarters) = $19,972 savings per year

reduced by 176 subs for the 3 months, $1,927.[3] Projecting these figures to the entire year, we find a saving of $38,600. (This assumes that the churn reduction experienced in these 3 months would be a permanent factor, month after month.)

There would also be a saving in change of service costs in this nonaddressable system. Assuming that a downgrade service call costs $11, the annual saving in service calls is $40,128.[4] The total annual saving in subscriber fees and service calls is $78,730.

With the significant reduction in churn through the guides, it would be safe to say that other retention activities (and expenditures) could be reduced. At churn levels as low as those reported for cable guide households (around 2%), it would be difficult to realize much benefit from retention advertising and other promotional efforts. One must also assume that a certain level of churn is inevitable as a transient population moves on and the downswings of economic fortunes hit some households. On the other hand, since some subscribers did not make much use of the guide, other types of retention activity are still important.

Note that the guides were most used and most appreciated by the one-pay and multipay households—those households with the greatest potential for churn.

There is some risk in taking these retention results at full value since the guide households knew they were participating in an experiment. They knew they would be periodically asked to evaluate the guide. Could this pique their interest in cable, temporarily, or discourage them from disconnecting or downgrading? Perhaps.

Of course, the guide cannot achieve its objectives if it is not perceived as convenient and useful. In fact, a poor guide could have a negative effect on the company image and, ultimately, on retention. Two-thirds wanted it to continue. Almost all thought the size was right and that the guide was useful.

Some factors that relate to retention could be influenced by the guide. Subscriber attitude is one such factor. In the study, attitudes towards the franchise were measured. Presumably, editorial content designed to build the company image and explain company functions would improve these attitudes. This did not prove to be the case. Apparently, this kind of guide content either received little readership, or its effect, over so short a period, was not measurable. Since the primary use of the guide is to find out about programs, the editorial material may

---

[3] .002 (reduction in churn) × 6,094 subs (October) = 12
.017 (reduction in churn) × 6,091 subs (November) = 104
.008 (reduction in churn) × 7,544 subs (December) = 60
176 × $10.95 = $1,927 × 4 (quarters) = $7,708 savings per year
[4] 280 basic only subs saved over 3 mo. × 4 quarters = 1,120
456 HBO only subs saved over 3 mo. × 4 quarters = 1,824
176 two-pay subs saved over 3 mo. × 4 quarters = 704
1,120 + 1,824 + 704 = 3,648 × $11 = $40,128

be marginally attractive. After all, cable subscribers are filling leisure time with television viewing. The attraction of viewing distracts from guide editorial. While it might be premature to abandon editorial efforts at company image building, the efforts here did not have a significant payoff.

It was also hypothesized that, if cable subscribers would only learn more about cable, they would have a greater appreciation. The guide, in both its editorial pages and program listings, could teach subscribers about cable—how it works and what it offers.

While the results, at Wave 1 as well as Wave 3, indicate that there is plenty for subscribers to learn, overall knowledge about cable did not increase over the 5 months of guide availability except for the two-pay subscribers, who made some slight improvement. These were the people who also made the greatest use of the guide. Presumably, the editorial content would carry the burden of informing subscribers about cable generally, which, as discussed above, probably has little impact.

Perhaps more important is the increase in knowledge of specific cable channels. One-fourth of the subscribers said they found out about channels they hadn't previously known about through the guide. However, the number of channels viewed did not increase over the trial period. The guide may have enhanced the value of the favored channels, but it did not seem to increase the breadth of viewing across the available channels.

Upgrades.   There is no direct evidence from this study that would suggest the guides encouraged upgrades. It is clearly possible that the regular exposure to program offerings on premium channels to which the household does not subscribe would stimulate interest in those channels. It is perhaps too much to expect that this interest would, of itself, increase penetration. But it is feasible that interest in the unavailable programs would create a more receptive environment for remarketing.

Therefore, it is worth considering the findings on the use of the guide. The guides were used, and the most important elements were the primetime grids and daily program listing.

The two-pay subscribers make the greatest use of the guides. In this cable system, they have nowhere to go. They cannot upgrade. The HBO-only subscribers used the guides somewhat less, and the basic subscribers least of all. If the guide is to improve the climate for upgrades, unfortunately, those households with the most potential for upgrade get the least exposure to the overall program listings.

Potential for producing revenue.   This study did not address directly the issue of using the guide to produce revenue. Had the guide included general advertising, we could have determined if the advertising had a positive value for the guide users, and whether or not the advertising interfered with its principal functions.

Recommendations. If the guides cost $.35 per copy and the mailing and handling cost is $.11, the annual cost of the guide mailed to all subscribers is $230,988.[5] The saving in subscription fees and service calls of $78,730 must be evaluated against this cost.

Looking at the three types of subscribers separately, the saving in subscription fees plus the saving in service calls costs for *basic-only* subscribers is $23,240, against the guide cost for basic-only subscribers of $139,579.[6] For *HBO-only* subs, the saving of $40,036 in service and subscriber fees compares to $91,128 in guide cost.[7] For *two-pay* subscribers, the subscription fee and service call saving is $15,452 against the guide cost of $39,456.[8] The guide cost for pay-only subscribers is $.50 per copy. Since the press run is less than half of the press run that includes basic subs, the subscriber cost would be higher.

It can be seen from these figures that the principal difference between the cost of guides and the savings from churn reduction (in subscriber fees and service calls) is with basic-only subscribers. The annual loss of $75,096 could be made up in subscriber fees for the guide. However, subscription to the guide should probably not be voluntary, since the subscribers most vulnerable to churn would probably opt not to take it.

To be even more specific, a per-subscriber cost of $3.50 per year (30 cents per month) would cover the cost of the guides to the pay subscribers (in addition to the assumed saving generated by churn reduction at the average level of the final three months of the guide test period).

It is important to note one change in the television information environment of this particular market. During the time period in which the experimental guide was tested, the local newspaper was distributed in the afternoon, and was likely to be available and used during evening viewing sessions. Despite this, the experimental guide became a primary information source for large proportions of cable subscribers, particularly the pay subscribers. Subsequently, that newspaper converted to a morning paper and is less likely to be as accessible in the evening to guide viewing. Thus, the expectation that subscribers would be even more reliant on the cable guide as a television reference may have additional credence; its use would likely be enhanced.

---

[5] 47,050 average copies per month × 12 mo. = 564,600 × .35 = $197,610

25,286 average basic only subs × .11 × 12 mo. = $33,378

Mailing charges to pay subs are not included, since the system specific guide replaces the current HBO/Cinemax guide with similar costs.

[6] 25,286 (average basic only) × 12 = 303,432 × .35 = $106,201 + $33,378 (mailing/handling) = $139,579 cost

1,120 × $11 = $12,320 (service) + $10,920 (sub. fees) = $23,240 saving

[7] 15,188 (average HBO only) × 12 = 182,256 × .50 = $91,128 cost

1,824 × $11 = $20,064 (service) + $19,972 (sub fees) = $40,036 saving

[8] 6,576 (average two-pay) × 12 = 78,912 × .50 = $39,456 cost.

704 × $11 = $7,744 (service) + 7,708 (sub fees) = $15,452 saving

Administrative costs of providing program information to the guide publisher and other liaison are ignored.

## CHAPTER 21

Conclusions
and a Research Agenda

Bradley S. Greenberg and Carrie Heeter

In early 1987, just under 50% of American households were receiving cable television and were expected to spend $12 billion for that privilege by the end of that year. In 1986, there were more than 42 million subscribers, of whom 23 million were subscribers to at least one pay channel. More was being spent on cable television subscriptions than on going out to the movies ($3.89 billion) or on video cassette rentals ($3.4 billion). Cable-only networks were credited with at least one-third of the viewing shares in cable households, at the expense primarily of the three major commercial broadcasting networks; the commercial networks' prime time viewing shares decreased from 89% in 1978–79 to 73% in 1985–86, and their daytime viewing shares had a similar drop (from 77% to 61%). And in 1987, the cable industry has projected that advertising revenues will exceed $1 billion for the first time.

These impressive statistics on the diffusion and utilization of cable television do not tell us much about individuals and their accommodation to this relatively new (for some) form of television. That, of course, has been the main thrust of this volume.

The more than 20 studies reported in this volume dealt with the viewing of cable television, in one form or another. Gathered from different, but not all, parts of the country and from a variety of cable systems, with different audience segments, the evidence cannot be considered national in any formal sense. Yet, where similar findings do accumulate across studies, they provide the basis for beginning generalizations. We set out to identify changing patterns in the utilization of television in an expanding channel environment brought on by the introduction and acceptance of cable television, confounded by remote control selectors, and altered by the diffusion of still another television technology, the videocassette recorder. We wanted to look at viewing patterns with certain preconceptions in mind, particularly those that dealt with an interest in how cable viewers cope with this changed viewing environment. For that reason, the studies go well beyond what cable subscribers choose to watch and instead focus most heavily on the choice processes available to them and how they exercise themselves in that effort.

From this massed territory of studies, it is time to draw back and to survey what has been uncovered. And then, perhaps, it is appropriate to assess what remains buried. Before analogies get the best of us, let us route ourselves through these data sets to look for common themes or sufficiently common findings. These can serve as the strongest markers of what has been found.

Several research issues guided this assemblage. To the extent we can encompass the studies' findings within these issues, they provide a convenient means of summarization. They also link back directly to the conceptualization provided in the opening chapters of this volume. The first issue is the extent to which *viewing is planned* and how it is planned; the second evolves from the kinds of *orienting searches* developed as individuals choose from among what is available; the third looks to the *allocation of viewing time*, and to some of the content choices made; and the fourth focuses on reevaluation or *channel-changing behaviors*. In addition, there were several tests of the choice process model; i.e., information was analyzed across and among these variables, and several profiles of different viewing groups were examined. So, these highlighted issues provide us with guidelines for the discussion and summary which follow.

Planned viewing.   Here the use of television guides to provide program planning information was the key variable. Determining whether viewers planned ahead for the week's viewing or even the night's viewing had little payoff and did not differentiate cable from noncable viewers. However, guide use of some kind was maximum in pay cable homes and minimal in noncable homes, and this differentiation in guide use among pay, basic, and broadcast homes held up in the most stringent analyses. Over time, however, and as a likely result of guide usage, preselection of programs became more common. Cable viewers were more likely to know what they would be watching, and this held for basic, one-pay, and multipay subscribers. Guide use itself did not increase over time; it was greater to begin with in cabled homes, and remained that way.

In the experimental study of guides—providing a system-specific guide where none had existed—there was a dramatic shift in which guide was used. Newspaper guide use dropped among pay subscribers, but the largest decrease was among magazine-type guides, which were replaced by the system-specific guide. By the time of distribution of the final (fifth) monthly edition of the experimental guide, its use was equivalent to that of any other magazine guide, including *TV Guide*. Again, maximum use was found among the multipay subscribers—they were more aware of the guide, wished it to continue, and rated it more useful.

Who plans? Parents make more use of guides than their children; however, parents who use guides and plan a given night's viewing are likely to have children who exhibit those same behaviors. Older viewers and those with higher incomes do more planning and more guide-checking. Females are more likely to know what they will watch before they turn the set on, while males do more channel checking at that time. Females make greater use of a guide before watching television, and they are more likely to plan for the week's viewing than males. Those with more-active social lives are less likely to know what they're going to watch before they turn the set on; television is a lesser portion of their leisure-time emphasis. Zappers are less likely to know what shows they'll watch before turning on the television, and are less likely to use guides or know the week's offerings.

Guides are used more so to check on movies and television series than on news or sports offerings. Preplanning occurs more for television and home radio use than for car radio use. But on an overall basis, preplanning is not a frequent viewing behavior; the best estimate is that some form of preplanning, primarily in the form of guide use, occurs about one-third of the time. Nonplanning then would seem to be a combination of habit strength, perceived or actual knowledge of what's available, or orienting search behaviors that resolve the immediate information needs for making a program choice.

Orienting searches.   A series of search typologies was developed in Chapter 2 along the dimensions of automatic–controlled, exhaustive–terminating, and elaborated–restrictive approaches to deciding what to watch after turning the set on. For the present purposes, these were operationalized most often in terms of whether the viewer searched through the available channels in numerical sequence (from 1, 2, 3 . . . n in order), in a skip numerical sequence (2, 5, 7, 8 . . . n in order), in nonnumerical sequence (3, 9, 4, 11, etc.), whether the viewer stopped at the first show that looked good, whether the viewer went through a subset of channels and returned to the best-looking one, or watched whatever happened to be on the set when the set was turned on. In addition, we determined the total number of available options that were searched through.

First, cable viewers do more in the way of orienting-search activity; they check more total channels, and they more likely check in some numerical sequence; of course, they have more to check than nonsubscribers. Nevertheless, the pattern continues among cable viewers, whereby orienting search activity is a significant discriminator between basic and pay subscribers, with the latter as more active searchers. However, over time within a season, less searching is found among each of the cable subscriber types.

Nonselective viewing was a composite index of two interrelated orienting search measures—stopping at the first show that looks good and watching whatever happens to be on. Having children (under 18) in the household stimulated that form of searching, whereas older viewers rejected such searches. Doing other things while watching TV enhanced nonselective viewing (more channel checking), as did having a more active social calendar.

Orienting searches for television exceeded those for either car or home radio

use. Older and more affluent viewers do less searching. More specific differences occur between the genders: males are more likely to stop at the first show
that looks good, are more likely to look through a set of channels and then go
back to the best looking show, and they are more likely to check channels out of
order. Both men and women check channels in numerical order to the same extent; that method is the more prevalent female option for orienting themselves to
television. Among children, the sex difference is consistent: young girls use this
method significantly more than young boys. The orienting search pattern most
commonly found within parent–child pairs is checking multiple channels and
then returning to the most favored one; that pattern of orienting search, coupled
with channel checking during programs, forms a primary pattern for parents and
their children in a canonical analysis.

Zappers do more searching of virtually all kinds: they check more channels,
they do so in numerical order more so, and they are more likely to check channels and stop at the first show that looks good.

Viewing-time indices.   Now that we have them watching something on television, how do they allocate their time and to what? The most intriguing finding
may be the identification of viewing-time allotments in strings and stretches
among cable households, time segments that are unlikely to exist, at least to the
same extent, in uncabled settings. However, we have not been able to verify the
prevalence or rate of string viewing (and several levels of stretches) in uncabled
homes. We must content ourselves here with advocating what appears to be
behavioral regularity in cable settings.

In the studies utilizing electronic measurement of household viewing, strings
of 1 to 4 minutes constituted 8% of the time spent viewing, ministretches of 5 to
14 minutes comprised another 10%, and longer viewing periods averaged 56
minutes (or an hour program). Subsequently, when stretches were split into
stretches and maxistretches, it was found that viewing segments over time
(specifically as the new season took hold) became longer. Maxistretches increased as a proportion of viewing at the expense of all the shorter segments,
indicating that a portion of the early season effort is sampling from among
shows, but not sticking with those that are disappointing, even for one entire
episode.

Nevertheless, young viewers, or at least households with children under 18
years of age, spent larger proportions of viewing time watching in short viewing
chunks, i.e., more strings and ministretches. Whether these youngsters will
elongate their viewing as they age, or whether the pattern they have learned at a
much younger age remains with them, becomes one of the challenging issues to
consider. Does viewing in spurts suggest an alternative programming style, a
style more akin to MTV than to hour-long dramas?

Relatively small sets of variables could account for substantial portions of the
variance associated with this phenomenon of viewing in varying amounts of
time. Typically, 25% of the variance in total viewing time, string viewing,

ministretch, and stretch viewing could be accounted for by some combination of these variables:

knowing the number of adult viewers in the household,
the prevalence of communication gadgets in the home,
the leisure time preferences (with sports preferences a positive influence
  and audio preferences a negative influence on viewing times),
education, and
heavy reading tendencies as negative influences.

One process characteristic consistently related to viewing in these time chunks was what was labeled the *more adventurous* viewer, the individual who more often claims not to know what will be watched before turning on the television, and who tends to watch more than one show at a time when watching. Not predictable to nearly the same extent was viewing in the longest time stretches—the "maxistretches" of 25 minutes or longer. Given the relative prominence of this viewing time segment across all individuals, differentiation is more difficult.

Cable viewers watch more television; likely, they watched more television before they had cable and the selectivity that went into subscription cannot be unconfounded from their prior state. But, from the experimental marketing study where a month of cable was offered free, viewing increased rapidly by about 45 minutes a day. This was not a perfectly controlled experiment, but one which more vigorously suggests that there is some initial viewing gain above and beyond what was already present before the cable came into the home or town. And consistently, pay subscribers spent more time with television than basic subscribers, perhaps to justify, or at least to test out, their additional investment.

So, how is the time spent with television distributed? Cable viewers spent more time with local or distant network stations than with cable-only stations. Pay subscribers gave more time to virtually all categories of cable-only channels, including text channels, than did basic subscribers.

With cable, the commercial networks' viewing share was diminished, falling below half of all viewing time in summer periods, but generally accounting for two-thirds to three-quarters of all viewing time—a sharp intrusion into what was once the sole province of the commercial networks, save for 3% to 4% of public television viewing. Nevertheless, another study shows that the network share increases substantially as the fall season's offerings are sorted out by the viewer. There are corresponding decreases in shares for PBS and independent stations as the fall season takes hold. Pay cable channels account for perhaps one-sixth of viewing time, while superstations, independents, and general satellite networks account for shares totalling in the 10% to 14% range. But the remaining 15 to 20 channels on typical 36-channel systems draw relatively insignificant viewing audiences and basically fill available channel space.

In the study of Saturday morning viewing tendencies in homes with and without children (under 18), there were no differences in total viewing time, but a sharp difference in where that viewing was allocated. For child households, it was predominantly to the networks (61%) and then to nonpay cable channels (22%), with pay channels attracting a most modest child audience during that time period (12%). For homes without children, the networks and nonpay cable channels each attracted one-third shares, with one-sixth more going to pay channels.

The parent–child correspondence in viewing tendencies remains intriguing. In noncabled homes, the parent–child relationship in usage of available channel options was virtually nonexistent; in cabled homes, there were significant relationships in the amount of time spent with 15 cable-only channels, including the pay channels available. This latter finding probably reflects the availability of the pay channels in the home, and the joint viewing of those channels' offerings.

News viewing from cable is trivial relative to the amount of cable news time available, but nontrivial as a portion of total news exposure exercised by an individual viewer. Cable-only channels provided nearly 25% of the information programming viewed per day. We found nearly 10 minutes per day going to news on cable channels, and less than 1 minute per day going to automated or text channels. This is deceiving, however, because four of every five viewers devoted no time to information content available on cable. Thus, those who do utilize these news resources do so for nearly 45 minutes per day. So the specialty cable channels with news content have an avid audience segment, and that is what cable intended to do in part, provide smaller segments of the viewing audience with its desired content.

Channel changing.   Reevaluation of one's television viewing decisions is a constant process. Perhaps we are not yet a nation of frenetic viewers, but we certainly are becoming more active in this process, given access to remote-control channel-changing devices in combination with a larger array of channel-viewing options from which to pick.

We still have no good count of just how much channel changing occurs. Recall that the measurement frequency in our two-way cable studies assessed channel changes on every minute of viewing. Doing so counted an average of 34 channel changes per household per day at this criterion. But it is a safe speculation that far more actual channel changes occur than would be counted checking only every minute, e.g., the 1- or 2-second scan of each of 20 to 30 channels by those who do elaborated or exhaustive searches as they start to watch television, and those who do the same thing at commercial breaks or between programs. Continuous measurement of channel changes is costly but technologically feasible. When someone with adequate funding decides that it would be valuable to know just how much occurs, when it occurs, to what channels and away from what channels, then more precise estimates of this phenomenon will be known, and less reliance need be placed on self-reported judgments of how much changing viewers do. For households with active channel changers, scanning most channels available three or four times in an evening of viewing, there will be several dozen recorded switches.

Channel changing occurs most often on the hour and on the half-hour, as would be expected. But the data also reveal that changing goes on throughout the typical television hour. Seldom in any single minute of the 60-minute hour does channel changing fall below 70% of the kind of distribution that would be found if all changes in the hour were evenly distributed across that hour. And, day after day, the change pattern in the same households looks remarkably similar, suggesting that the habit strength of channel changing may be as important a factor as reactions to specific content.

We also found maximum channel changing at the beginning of the fall season, minimum changing in the summer, with mid-fall changing behavior regressing to the summer level (and no measurements in mid-winter for comparison). More channel changing occurs in homes with children than without, and more channel changing occurs with cable-only channels than with commercial broadcast networks. This latter behavior is not a function of the greater number of cable channels. By comparison with viewing time share, channel changing was more prevalent in cable-only networks; for example, network channels had 44% of the viewing and 36% of the changes, whereas nonpay cable channels had 23% of the viewing and 37% of the changes. This was found even more extensively in homes with children, where, on Saturday mornings, 61% of the viewing went to the commercial networks and 22% to nonpay cable channels, in comparison with 45% and 31% of the changes, respectively.

So, who changes? A major point to be made is that approximately one-third of the adults interviewed across studies indicate that they do no substantial zapping. On the other hand, males are far more active changers than women; men change more between shows, during shows, during ads, and they are more likely concurrent multiple-show viewers. Women more often indicate they watch shows in their entirety, and more often complain that someone is changing channels whom they wish would not. The younger generation is more active than the older: this is reflected in differences both between children and their parents, and younger adults contrasted with older adults. Also note that there is a substantial correlation between parents and their children in this viewing-style attribute. Parents and children show related tendencies to change channels—at all possible time segments—and to watch more than one show at a time. Differences are minimal between income and education groupings, between marrieds and otherwise, and even by household size. Furthermore, there is no apparent total viewing time difference between zappers and nonzappers; it is the way they watch (and in that sense, what they see) rather than how much they watch.

Zappers carry their channel-changing behavior into a variety of related viewing style manifestations: they generally indicate they pay less attention to television, use it more as background, are less likely to watch whole shows, and more likely to watch multiple shows concurrently.

Cable subscribers do more reevaluation than nonsubscribers, both in terms of general channel-changing propensities and in terms of multiple show viewing. No zapping difference was found between basic and pay subscribers, but pay subs were more prone to attempt to maximize their investment by greater watching of two shows at the same time. Over time, cable viewers do not evidence significant changes in their general channel-changing behavior. However, there is a cyclical pattern. Viewers tend to watch entire shows more often and do less multiple show viewing later in a season. If one separates out multipay subscribers—typically the most frenetic channel changers—they reduce the level of that behavior over time.

Channel changing peaks in prime time, with most changes coming from other cable networks, and least from commercial networks. The prediction of total channel changes was best accomplished by knowledge of the number of adults in the household, the leisure preferences in the household (positive for sports activity and negative for audio activities), and the reading tendencies (negative relationship).

Channel repertoire. Early in this volume, two variables were considered as outcome variables from the viewing choice process. These included channel familiarity (or how many channels available the viewer was familiar with) and channel repertoire (the number of channels used on a regular basis, e.g., once a week or more). The second of these was examined in several studies, the first less so.

Most often, the studies were conducted in 36-channel systems. Typically, a linear relationship was found for both familiarity and repertoire as one moved from broadcast-only households to basic cable homes to single-pay and to multipay. This distinction held up in multiple regression analyses as well as bivariate analyses. Stronger channel repertoire meant three to five more channels among the multipay subscribers as compared to the broadcast-only homes.

In one study, as an example, in the Lansing, Michigan, Continental Cable system, the broadcast households' average channel repertoire was four of the eight available off-air channels. For basic cable subscribers, it was 11; for 1-pay, it was 12; and for 2-pay or more, it was 13. Over a 6-month period, these averages did not vary. In a different study, it did not vary over a 6-week period. And the television channel repertoire was larger than that for radio (in a community with 16 possible radio stations).

In the parent–child comparison in another community, the parent's repertoire was nine channels with cable and four without; their children averaged 11 channels in their repertoire, if they had cable, and five without. Furthermore, the first root (33% of the variance) in the canonical correlation analysis of parent–child pairs was that of households with a large channel repertoire. Elsewhere, focusing solely on Saturday morning viewing, the overall channel repertoire used was smaller than on a week-long basis, and households with children had larger channel repertoires than households without children. No differences in channel repertoire size emerged by gender among either children or parents.

Age, however, was a significant correlate of the number of favorite stations identified by the respondents, as was the number of hours that television was viewed. However, those with more active social lives were likely to have fewer favorite television channels.

Models. Finally, let us return to the model of cableviewing for which this mountain of data was mapped and scaled. In doing so, the linkages among these behaviors and postures have been examined:

PLANNING—
    ORIENTING SEARCHES—
      REEVALUATION—
        CHANNEL REPERTOIRE.

In several studies, what came to be the full model was available for analysis; in others, portions were examined. We are comfortable drawing these conclusions:

1. Planning one's viewing (to include advance consideration of an evening's viewing by reading the television log in the morning paper, thinking about what is typically available on a given week night, or deliberate consideration of a more elaborate television guide magazine) has a modest negative relationship with the amount of orienting search that is found. That relationship generally is at about the .20 correlational level, although, with Playboy subscribers, this was obtained only with how they made choices on the Playboy channel and not in their general viewing orientation. Planned viewing on an absolute level is not a majority behavior; although not specifically established in these studies, it is likely that much television watching is impulse watching, akin to impulse buying at the supermarket.

2. In similar fashion, advance planning of television activity is negatively correlated with subsequent reevaluation of program choices; i.e., the more thought put into deciding what to watch, the less likely that choice is reversed, with correlations in the same range as noted above.

3. Orienting searches have fit into some intriguing typologies. However, rather than the kind of orienting search proving to be very useful yet, it is more the size of the search, the number of channels involved, that has been most often used in testing component parts of the model. This appears to an analog of effort from the viewer at this stage in the process, how hard he or she is trying to find something. The extent of the orienting search has a consistent and positive relationship with the amount of subsequent reevaluation that occurs, typically in the .30–.40 range. These are of course redundant behaviors in the sense that channel changers are channel changers, regardless of whether it is when they initially sit down to watch television or after they have been watching for awhile. On both occasions, those who change channels continue to do so.

4. Reevaluation and orienting search magnitude put the viewer in contact with more channels. It is not unexpected then, that both of these aspects of the

viewing process are tied to the regularity with which the viewer espouses favorite channels. Hence, the viewer's channel repertoire, as well as channel familiarity, is linked more directly to the activism found in those behaviors than in planning, which has virtually a nil relationship with those outcome variables.

Some other findings bear on the heuristic utility of this model. For example, testing it against radio use (at home and away) and finding the relationships not strikingly different between these media provides more generalization capability than originally anticipated. We think acceptable parallel behaviors were defined, and the results are supportive; in communities with even greater numbers of radio station options, we would anticipate an even stronger likelihood of finding people behaving in these ways.

The fact that model components were supported in cable contexts and not in broadcast-only contexts enhances the notion that what we are examining is peculiar if not unique to the multichannel environment. Having cable (rather than not having cable) showed a consistent relationship with time spent viewing, with use of a television guide, and with the extensiveness of one's channel repertoire. It was certainly anticipated that having cable should also be predictive of channel-changing behaviors, and it is, but yet to be sorted out is the extent to which channel changing can be accounted for by remote control capability and by the extensiveness of channel options.

Finally, in this synthesis of the findings, the parallelism of cable-viewing behaviors between parents and their children and the consistent differences in these behaviors across generations merits comment. How can we argue effectively that both conditions can co-exist? Not overly difficult, it turns out. The fact that parents and their offspring claim for themselves behaviors that are correlated masks the absolute differences in magnitude of the behaviors claimed by each. So, parents who do more channel changing have children who do more channel changing, and the paired behaviors are correlated; however, at the same time, the children do more of that behavior on an absolute basis, and thus significant differences emerge between the paired members. What cannot be determined from these kinds of data is who is influencing whom. It seems as reasonable to argue that with these new technology "toys" the child is the role model, is more facile in channel changing, and the parent may be the imitator. Such an argument could be premised on the notion that, for the child, use of cable and use of remote control devices is more nearly a lifetime habit, that no prior learning of earlier systems need be unlearned. That is as arguable as the more traditional notion that the child models the parent.

Research Agenda

The last paragraph begins the research agenda, but let us approach it a bit more systematically. If we could have our druthers, several additional major studies would now be undertaken to advance examination of the ideas developed here.

The first would focus on the generalizability of the concepts and findings. This would consist of a national-sample examination of the cable-viewing process, in which a stratification of cable subscribers would segment the number of channel options in the system, and for which a parallel national sample of uncabled viewers, stratified as to the magnitude of off-air signals available, would provide some semblance of a control group or useful comparison group. Among the former, it would be valuable to have viewing-style data from subscribers in 12-, 24-, 36-, and 54-channel systems, for the basic proposition has been that, as the number of viewing options increase, the decision process requires greater effort in decision making. Further, the presence or absence of remote control, in a national sample, could be separated from the impacts of cable, and impacts of different types of channel selectors could be determined. This effort should be manifested in alternative uses of the planning, orienting searches, reevaluation, etc. mechanisms described. To date, we have no direct comparisons among such systems. At the same time, there are media environments without cable which have alternative numbers of off-air signals, e.g., 2, 4, or 8 broadcast television channels. That is why it is proposed that a second sample of uncabled viewers be included for comparison.

The second study would be a more direct examination of what happens when cable comes to town. To date, studies of the impact of the introduction of cable television have been post hoc. Researchers have typically come into a community sometime after cable has diffused and made one of two approaches: (a) to ask people whether they are now using television the same or differently as before they had cable, invoking all of the problems of memory recall and desirability of response, e.g., who wants to admit they are watching television more now than before; and (b) comparing those with cable with those without cable and making statements about the impact of cable, invoking all of the problems of selectivity in who chose to obtain cable and who chose to reject it. So, it is argued that this second study be conducted in either of two designs. The first is probably the more executable; it would consist of identifying a community about to be cabled, and following a sample from the precable period through the early diffusion and adoption period up until at least cable has been in use for a year (or not in use among households which reject it). This calls for four or five assessments among a panel of households, but would permit determination of cable's impact, with parallel data from rejectors. Another option would be to obtain comparison data from a second, comparable community still uncabled.

The second study design would be more experimental. Again, the need for a community about to be cabled is a starting point. Then, to a random collection of households, offer cable for free, and withhold any such offer from an equivalent random group. Assuming most would take it on a free trial basis, as demonstrated in the experiment reported in Chapter 19, one could at least determine short-term results. This of course makes for poor marketing practices, and would probably not be tolerated by the local franchise-awarding body. Another alternative would be to make the same trial offer within an area where cable becomes available—and before it is available throughout the community; then perhaps an appropriate control group can be obtained from among homes where the cable has not yet passed. The final option would be to provide the researchers with sufficient funding to obtain a cable franchise (for the length of the research project), so that these study arrangements can be facilitated. Then, from profits obtained in resale of the franchise, additional research could be undertaken. (Of course, the moon could turn blue, couldn't it?)

We have been dependent for most of these studies on individual recall of their behaviors, orientations, and attitudes. Only in the two-way cable studies could we obtain electronic measurement of viewing behaviors, and those viewing measures were more imprecise than are required for second-by-second channel changing assessment. The kind of orienting search that takes a would-be viewer through most or all of the channel options in a 36-channel system is probably accomplished in 2 minutes or less. Thus, we do not have precise measures of the channel-changing extensiveness believed to exist; we have relative answers and they no longer seem sufficient. This is so because contemporary technology in audience assessment can today do much more readily what was basically a novel attempt 3 years ago.

So we would opt for the resources (and interest) from the Nielsens, AGBs, and Arbitrons to make such measurement available, at least long enough and broadly enough so that the parameters of channel changing can be appropriately examined. It would serve well the interests of clients and their advertising agencies to have more precise data on commercial zapping than any have chosen to provide to date in this ongoing argument. In zapping a commercial, for example, to what extent is the movement made to another commercial, rather than to program content? Or to what extent is movement away from a commercial only a momentary diversion, followed by return to that commercial before its closing? Or does the movement away from a commercial usually occur only after the viewer has already identified the product, brand, etc.? All these good market questions can be answered and at the same time provide researchers with the kind of precise channel-changing measures that would aid understanding of the viewing style associated with decisions to move across channel options.

Turning now to more subtle issues emerging from this research collection, several evolve around *whose viewing style is being examined*. Electronic data of the kind obtained from Temple Terrace provide household information; in asking above for more precise electronic measures, the reference is to individual viewing data. Household data may have useful commercial purposes, but the focus of interest in these studies has been to examine and explain individual viewing decision-making styles. Once we can do that however, we are faced with the valid criticism that much viewing occurs in a social rather than a solo context. Although one or two studies refer to viewing with others, and particularly to the possibility that some viewers become upset when others in their midst exhibit their channel changing skills (or deficiencies?), there has been no extended examination of group viewing. One would certainly expect viewing in a group to provide inhibitions or constraints on viewing-style behaviors that would be more prevalent when viewing alone. On the other hand, the larger the group, the more likely someone in the group is a zapper. But earlier than the issue of changing channels would be the decision process to arrive at a consensual channel/program choice or the extent to which that decision is dictatorial rather than consensual, and by whom. Who controls the channel selector, the remote control device and how is that control exercised?

Another pertinent issue is the extent to which what are being characterized here as viewing-style behaviors are enduring ones or transitory. Can individual data on viewing styles be cast into a typological framework, so that we have planners and zappers and exhaustive searchers and guide users, and so on? Have we become habituated in our viewing style so that we are largely impervious to change, at least for internally motivated reasons? It is apparent that, when cable systems introduce new channel options, or when they add an on-screen cable television guide, or when they reassign channel numbers, we become aware of and adapt to those changes—all externally driven? But from within our own choice framework, is it likely that we can shift from a nonselective television viewing posture to one of considerable advance planning? Only over-time studies can answer these sorts of questions.

For the moment, let us argue that our styles are persistent—not necessarily inflexible, but that we tend to maintain one style by and large over others. Then, how did it come to be? From what individual or environmental characteristics has that viewing style mode been derived? What personality or social attributes prod us into adopting and maintaining a particular style? For example, clear evidence is provided that men are more active channel changers than women, and that even young boys exhibit this tendency over young girls. Why? Does one seek an explanation in some "power" conception, that having control over the device is a form of male dominance in the household? Or is it that males have a shorter attention span, and so move sooner to alternative attention-getting passages?

The parent–child similarities in style, and differences in amount, of behavior provide a useful example. Do young children develop patterns of viewing television that consistently remain with them? Here, we would argue that is likely to be the case and increasingly so, as more and more homes become better equipped with remote control devices, cable television multichannel options, video-cassette recorders, personal computers, etc. For the child growing up in such a media-rich environment must seek ways to maximize efficient use of all that is available. Efficiency requires standardization or routinization in approaching such an array of options. And the child has never known it otherwise. There is a generation of 15-year-olds for whom cable television has been the standard since birth; there is a generation for whom the VCR will be as integral a part of the available television in the home as the commercial broadcasting networks are. The outcome of this reasoning, if substantiated, is that the child will develop a relatively rigid behavior pattern in approaching television. But this does not tell us whether to expect the child to be a planner or a plodder, a zapper or a couch potato; for those predictions we must look elsewhere. On the other hand, the typical parent must unlearn former viewing habits once cable, VCR, pay channels, and other television innovations are added to the household. And that is why one can argue that the parent is as prone to learn from the child how to handle all of this as the child is to learn from the parent; actually, a reciprocal learning expectation may be anticipated, by which the parent teaches the child what traditional television has consisted of, and the child provides the parent with a model of how to handle multiple television treats concurrently.

Equally interesting is whether a particular television viewing style is characteristic primarily of television viewing or whether it mirrors a more general life style. Can it be that those with more active leisure lives, whose disposition of available leisure time is across more activities, are individuals who carry that disposition with them into the viewing situation? Or might viewing style relate to work style? Do those with hectic schedules, messy desks, or messy houses also seek chaos in their media experiences? Are the viewing styles we have been attempting to categorize primarily reflections of particular lifestyle variants? To the extent there is any, research which has tried to probe into lifestyle characteristics across these kinds of behaviors ought be examined, and parallels determined.

Although the argument for enduring television viewing styles appears to have merit, there is evidence to the contrary. Recall that, in Chapter 13, viewing styles for different kinds of television content were examined in a preliminary fashion. Sports programs, movies, news, and weekly series were the objects of inquiry, and channel-changing behaviors were reported to vary among these different kinds of content. Similarly, in Chapter 17, where Playboy Channel subscribers characterized both their general viewing and their viewing on the Playboy Channel, there were reported differences in how often they watched complete shows, checked other options, and so on, between general proclivities and their behavior with the Playboy Channel. Thus, although viewers may be characterized as one kind of viewer or another in some general sense, it is likely that content attributes also influence how we watch. The "down time" in sporting events may induce more frequent checking of program options; the steady flow of events in a news program may minimize change behaviors. Planning to watch news programs which appear at constant times is unlikely to require guidance. Planning to watch particular movies may require considerably more study of the movie schedule for the week or the month. The desire to see news does not demand a full search of the channels, if the known repertoire includes those channels with regularly scheduled newscasts. The desire to watch re-runs of *M*A*S*H* which may be available on a given day on three or four cable channels at three or four different times may well lead to an elaborated repertoire search, or the systematic examination of an available television guide.

Program attributes may impact on the viewing process as well. Many cable shows offer more variety in time segments than 30- and 60-minute chunks. MTV segments run 7 or 8 minutes; CNN segments vary considerably; text channels depend on the viewer's reading skills. These are a few of the time alternatives permitting or encouraging the viewer to alter specific viewing style behaviors to better fit the format of the program.

Finally, the specificity of the object being studied requires refinement. Asking individuals to characterize how they generally approach a situation may lead to socially desirable responses, or to some kind of internal averaging that does not do justice to the individual variance involved. Take as an example the question of zapping and the zapping of commercials in particular, whether within or between shows. As stated earlier, the best measures should come from electronic assessment of such changes, provided there is adequate and precise knowledge of who is doing the zapping. In the meantime, it is likely that researchers will rely on verbal reports of viewing style behaviors, and perhaps even the behavior of others with whom respondents are viewing.

Confining the issue to the viewer alone, and recognizing that zapping may vary to some unknown extent across program types and program content, just how does one get at the viewer's commercial zapping propensities? As indicated, the most common method used in this volume has been to ask for general tendencies, e.g., "When commercials come on, how often do you change channels?" Whether the response options range from "almost all the time" to "almost never," or from "very often" to "not at all," the viewer is suddenly to come up with an average tendency to a behavior unlikely to have been considered or evaluated prior to the interview at hand. There is little doubt that the viewer is accurately characterizing a general tendency, but there may be considerable doubt about the precision of the estimate; a better precision is necessary to satisfy issues related to loss of viewers, loss of advertising exposures, etc.

An alternative strategy, at least to gain some better intuitive sense for the process, would be to invite respondents to consider in greater detail their most recent viewing session, and to ask them to reconstruct their zapping tendencies during that session. For example, a sample of viewers obtained during the middle of the afternoon would yield viewers who had just watched some soap operas and/or game shows. To ask them to recall if they did any commercial zapping (or zapping of other content by changing channels) would tap memories that should not yet have faded as would occur with more delayed interviewing. Coincidental interviewing might well be best for the memory problem; interviewing about yesterday or last night's programs would be helpful; both would aid in determining how characteristic the general depictions offered by viewers are.

At some early stage in the research program developed and reported in this book, still another tool was created that might have helped with this particular issue. A piece of furniture was constructed (the CARMA device: Cable Audience Research Monitoring Acolyte) that contained a computer and VCR, and upon which sat a television set. When the viewer turned on the TV, it activated the VCR and the computer, which then recorded what was viewed and when it was viewed. The output revealed all viewing changes made. This prototype device (only one was constructed) was moved among some different homes for purposes of determining its durability, its reliability of operation, and the difficulties encountered in calling each day to remove the tape. The logic then would be to debrief the viewers as soon as possible with regard to whatever information one might wish. For the present zapping problem, one could re-show the tape or samples of it in some fashion, stop at each channel change, and ask why that particular change had occurred. This of course aids recall, but also provides some quite specific information to examine.

More recently, a parallel effort has been developed in England under the sponsorship of the Independent Broadcasting Authority, whereby the piece of furniture also contains a video camera; the output then provides a verification of viewership, as well as of channel changing. The resultant tape contains on its split screen both the program content being viewed and the scene of the room's occupants. Both this technique and that experimented with earlier are very tedious data-gathering strategies and are expensive to manage. Nevertheless for those who wish to quarrel with or supplement the kinds of evidence otherwise available, typically the generalized self-reports indicated, alternatives are available to address the problems quite directly.

Finally, cable viewing needs to be examined in the context of the changing home-media environment. With VCR penetration at 39% and rising, the cable industry is beginning to look for ways that cable and VCRs can complement each other rather than compete. The complexity and variety of configurations involved in hooking up a VCR to be able to record cable programs appears to be a major problem. A recent industry study reports that many cable subscribers who own VCRs don't have them wired to be able to record off cable. Research linking cable-viewing style with the variety of uses of VCRs will better document active viewing. The choice model should also be applied to VCR use.

For all questions posed or suggested in these concluding remarks, as well as those addressed in separate chapters (or not addressed by these remarks at all), additional research attention is demanded.

What began for us as a curiosity about how individuals process information and make decisions in a multichannel television-viewing environment has burgeoned. Cable systems expanded from 24 to 36 to 54 channels, and increased their penetration from 35% to 50% of American homes during the period of these investigations; video-cassette recorders doubled their penetration of American homes during the same time frame to more than 50%; another TV set was added to each home, on average; remote control devices both for cable television use and broadcast sets have become commonplace. So the viewing environment continues to change and to demand new decisions from the viewer. There is no demonstrable end to those changes evident, given the advent of compact disc for improved audio, interactive videodisc for home instructional purposes, and personal computers linked by cable or telephone to the television set for networking and information supplies, among other telecommunication devices likely to further intrude on the home. Although we may say "Welcome" to all of these, their impacts on the individual, the family, and our social and work groups stimulate a cornucopia of questions. Seeking answers will be demanding; hopefully, the material presented in this volume will serve to engage others in that search.

References

Agostino, D. (1980). Cable television's impact on the audience of public television. *Journal of Broadcasting, 24* (3), 347-363.

Anderson, D., Alwitt, L., Lorch, E., & Levin, S. (1972). Watching children watch television. In G. A. Hale & M. Lewis (Eds.), *Attention and cognitive development* (pp. 331-363). New York: Plenum Press.

Anderson, D., & Levin, S. (1976). Young children's attention to *Sesame Street. Child Development, 47*, 806-811.

Arbitron. (1981). Do we need a 'cable diary?'. *Beyond the Ratings, 4* (2), 1.

Arbitron Ratings. (1983). *Two-way cable diary test: A summary of findings.* Laurel, MD: Author.

Arbitron Survey: Cable reaches 40 percent. (1984, June 11). *Broadcasting,* p. 13.

Atkin, C., Greenberg, B., Korzenny, F., & McDermott, S. (1978). Selective exposure to televised violence. *Journal of Broadcasting, 22* (1), 47-61.

Baldwin, T., Abel, J. & Ducey, R. (1984). The media environment's study, National Science Foundation Report, Washington, D.C.

Banks, S. (1980). Children's television viewing behavior. *Journal of Marketing, 44*, 48-55.

Baran, S. (1976a). How TV and film portrayals affect sexual satisfaction in college students. *Journalism Quarterly, 53* (3), 468-473.

Baran, S. (1976b). Sex on TV and adolescent sexual self-image. *Journal of Broadcasting, 20* (1), 61-68.

Barnes, J., & Kelloway, K. (1978). *Cable television viewership: An examination of innovative behavior.* Working paper 78-14, Memorial University of Newfoundland.

Becker, L., Dunwoody, S., & Rafaeli, S. (1983). Cable's impact on use of other news media. *Journal of Broadcasting, 27* (2), 127-140.

Belk, R. W. (1974). An exploratory assessment of situational effects in buyer behavior. *Journal of Marketing Research, 2*, 156-163.

Belk, R. W. (1975). Situational variables and consumer behavior. *Journal of Consumer Research, 2*, 157-164.

Berlyne, D. E. (1960). *Conflict, arousal and curiosity,* New York: McGraw-Hill.

Bezzini, J., & Desmond, R. J. (1982). *Adoption processes of cable television.* Unpublished manuscript, University of Hartford.

Bowman, G., & Farley, J. (1972). TV viewing: Application of a formal choice model. *Applied Economics, 4*, 245-259.

Brenner, S., & Levy, J., with Ruppel, F. (1982, February). *UHF viewing and television channel selector type.* UHF Comparability Task Force Report, Office of Plans and Policy, Federal Communications Commission.

Bruno, A. (1973). The network factor in TV viewing. *Journal of Advertising and Marketing Research, 13* (5), 33-39.

Buerkel-Rothfuss, N., & Mayes, S. (1981). Soap opera viewing: The cultivation effect. *Journal of Communication, 31* (1), 108-115.

Cable/Video Research Center (a division of Opinion Research Corporation). (1983, April). Segmentation study of the Urban/Suburban cable television market. Paper prepared for the National Cable Television Association. Princeton, NJ.

Cannon, H., & Merz, G. R. (1980). A new role for psychographics in media selection. *Journal of Advertising, 9* (2), 33-36.

Chaffee, S., McLeod, J., & Atkin, C. (1971). Parental influences on adolescent media use. *American Behavioral Scientist.*

Chen, M. (1984). Computers in the lives of our children: Looking back on a generation of television research. In R. Rice and Associates (Eds.), *The new media: Communication, research and technology* (pp. 269-286). Beverly Hills, CA: Sage.

Chen, M., & Paisley, W. (1983, November). *Children and interactive media: Exploring the effects of the second electronic revolution.* Paper presented to Speech Communication Association, Washington, DC.

Collins, J., Reagan, J., & Abel, J. (1983). Predicting cable subscribership: Local factors. *Journal of Broadcasting, 27,* 177-183.

Cronback, L. J. (1951). Coefficient alpha and the internal structure of tests. *Psychometrika, 16* (3), 297-334.

DeFleur, M. L., & Ball-Rokeach, S. J. (1982). *Theories of Mass Communication* (4th ed.). New York: Longman.

de Sola Pool, I. (1983). What ferment?: A challenge for empirical research. *Journal of Communication,* (Ferment in the Field issue) *33,* 258-261.

Donnelley Marketing Information Services & Simmons Market Research Bureau, Inc. (1983). The marketing resource of the 80's: ClusterPlus.

Donohue, L., Palmgren, P., & Duncan, J. (1980). An activation model of information exposure. *Communication Monographs, 47,* 295-303.

Dorr, A., Graves, S., & Phelps, E. (1980). Television literacy for young children. *Journal of Communication, 30,* 71-83.

Ducey, R., Krugman, D., & Eckrich, D. (1983). Predicting market segments in the cable industry: The basic and pay subscribers. *Journal of Broadcasting, 27* (2), 155-161.

Eckblad, G. (1963). The attractiveness of uncertainty. *Scandinavian Journal of Psychology, 4,* 1-13.

Forkan, J. P. (1981, September 14). Things upbeat at Music TV. *Advertising Age,* p. 71.

Frank, R., Becknell, J., & Cloasky, J. (1971). Television program types. *Journal of Marketing Research, 8,* 204-211.

Frank, R. E., & Greenberg, M. G. (1979, October). Zooming in on TV audiences. *Psychology Today,* p. 92 + .

Frank, R. E., Massy, W. F., & Morrison, D. G. (1965). Bias in multiple discriminant analysis. *Journal of Marketing Research, 2,* 250-258.

Gelman, M. (1983, October 6). Cable news fails to harm local newscast viewership. *Electronic Media,* p. 18.

Gensch, D., & Ranganathan, B. (1974). Evaluation of television program content for the purpose of promotional segmentation. *Journal of Marketing Research, 11,* 390-398.

Gensch, D., & Shaman, P. (1980). Models of competitive television ratings. *Journal of Marketing Research, 17,* 307-315.

Gilbert, D. (1983). Sales management workshop. Perrysburg, OH: Continental Cablevision, Inc.

Goodhart, G. J., Ehrenberg, A. S. C., & Collins, M. A. (1975). *The television audience: Patterns of viewing.* Lexington, MA: Lexington Books.

Greeno, J. (1976). Indefinite goals in well-structured problems. *Psychological Review, 83* (6), 419-491.

Grotta, G., & Newsom, D. (1983, Winter). How does cable television in the home relate to other media use patterns? *Journalism Quarterly,* 558-591, 609.

Heeter, C. (1988). Implications of new interactive technologies for conceptualizing communication. In J. Salvaggio & J. Bryant (Eds.), *Media use in the information age: Emerging patterns of adoption and consumer use.* Hillsdale, NJ: Ehrlbaum.

Henke, L., Donohue, T., Cook, C., & Cheung, D. (1983, May). *The impact of cable on traditional television news viewing habits.* Paper presented at the International Communication Association convention, Dallas.

Hill, D., & Dyer, J. (1981, Winter). Extent of diversion to newscasts from distant stations by cable viewers. *Journalism Quarterly,* 552-555.

Hornik, J. J. (1982). Situational effects on the consumption of time. *Journal of Marketing, 46,* 44-45.

Husson, W. (1982, May). *ARIMA models of the attention patterns to television of 3- and 6-year-old children.* Paper presented at the International Communication Association convention, Boston.

International Thomson Communications, Inc. (1985). *Cablefile/85.* Denver, CO: Author.

Jeffres, L. (1978). Cable TV and interest maximization. *Journalism Quarterly, 55,* 149-154.

Kaplan, S. J. (1978, Spring). The impact of cable television services on the use of competing media. *Journal of Broadcasting,* pp. 155-165.

Katz, W. (1982/1983). TV viewer fragmentation from cable TV. *Journal of Advertising Research, 22* (6), 27-30.

Kerkman, D., Wright, J., Huston, A., Rice, M., & Bremer, M. (1983, May). *Preschoolers who get cable TV: Family patterns, media orientations, and television use.* Paper presented at the International Communication Association convention, Dallas.

King, C. W., & Summers, J. O. (1971). Attitudes and media exposure. *Journal of Advertising Research, 2* (1), 26-32.

Kozielicki, J. (1981). *Psychological decision theory,* Warsaw, Poland: PWN—Polish Scientific Publishers.

Krugman, D. M., & Eckrich, D. (1982). Differences between cable and pay-cable audiences. *Journal of Advertising Research, 22,* 23-29.

Krugman, D., Ducey, R., & Eckrich, D. (1983). Market composition and cable television use. Unpublished Report, Michigan State University.

Krull, R., & Watt, J. (1975, November). *Television program complexity and ratings.* Paper presented at MAPOR, Itasca, Illinois.

Lefkowitz, H., Eron, L., Walder, L., & Huesmann, L. R. (1972). Television violence and child aggression: A follow-up study. In G. A. Comstock, E. Rubinstein, & J. Murray (Eds.), *Television and social behavior (Vol. 3): Television and adolescent aggressiveness.* Washington, DC: U.S. Government Printing Office.

LeRoy, D., & LeRoy, J. (1983). *The impact of the cable television industry on public television.* Television Ratings Analysis Consortium Report. Pacific Mountain Network.

Lloyd-Kolkin, D., Wheeler, P., & Strand, T. (1980). Developing a curriculum for teenagers. *Journal of Communication, 30*, 119-125.

Lumpkin, J. R., & Darden, W. R. (1982). Relating television preference viewing to shopping orientations, life styles, and demographics: The examination of perceptual and preference dimensions of television programming. *Journal of Advertising, 2* (4), 56-67.

Lyle, J., & Hoffman, H. (1972). Explorations in patterns of television viewing by preschool-age children. In E. Rubinstein, G. Comstock, & J. Murray (Eds.), *Television in daily life: Patterns of use* (pp. 257-271). Washington, DC: U.S. Department of HEW, NIMH.

Maddi, S. (1968). The pursuit of consistency and variety. In R. Abelson (Ed.), *Theories of cognitive consistency*. Chicago, IL: Rand McNally.

Market Opinion Research. (1987, Winter). TV zipping and zapping: Boom or bust for advertisers? *Media, 1*, 4.

Morrison, D. G. (1969). On the interpretation of discriminant analysis. *Journal of Marketing Research, 6*, 156-163.

MTV fights back. (1984, August 27). *Broadcasting*, p. 10.

Munsinger, H., & Kesson, W. (1964). Uncertainty, structure and reference. *Psychological Monographs, 78* (9).

New direct mail boosts Tampa cable sales by 50%. (1985, January). *Cable Marketing*, p. 10.

Nielsen, A. C. (1983a). *Channel switching in prime time. Report to NTI client meetings.* Northbrook, IL: A. C. Nielsen Co.

Nielsen, A. C. (1983b). *DMA test market profiles* (pp. 206-207) Northbrook, IL: A. C. Nielsen Co.

Ogilvy & Mather. (1987). Flipping, zapping and zipping—is anyone out there still watching commercials? *Changing Media, 7* (1), 1-8.

Owen, B., Beebe, J., & Manning, W. (1974). *Television economics.* Lexington, MA: D.C. Heath.

Parker, E., & Dunn, D. (1972, June 30). Information technology: Its social potential. *Science*, p. 176.

Pearson, P. (1970). Relationships between global and specified measures of novelty seeking. *Journal of Consulting and Clinical Psychology, 34*, 199-204.

Peterson, R. A. (1972). Psychographics and media exposure. *Journal of Advertising Research, 12* (3), 17-20.

Press, S. J. (1972). *Applied Multivariate Analysis*, New York: Holt, Rinehart and Winston.

Reagan, J. (1982). *Effects of cable television on news use.* Unpublished manuscript, University of Michigan.

Reymer & Gersin Associates, (1982). Winning over non-subscribers in three cable systems. National Cable Television Association, Report.

Rogers, E. & Chaffee, S. (1983). Communication as an academic discipline. *Journal of Communication* (Ferment in the Field issue). *33*, 18-30.

Rothe, J., Harvey, M., & Michael, G. (1983). The impact of cable television on subscriber and non-subscriber behavior. *Journal of Advertising Research, 23* (4), 15-23.

Rubin, A. (1984). Ritualized and instrumental uses of television. *Journal of Communication, 34* (3), 67-77.

Russell, S. (1984, November 5). Marketing the options: Probing effective approaches to marketing cable. *CableVision*, p. 56.

Scherer, T. M. (1970). *Industrial market structure and economic performance*. Chicago IL: Rand McNally.

Schramm, W. (1983). The unique perspective of communication: A retrospective. *Journal of Communication* (Ferment in the Field issue), *33*, 6-17.

Sears, D., & Freedman, J. (1974). Selective exposure to information: A critical review. In W. Schramm & D. Roberts (Eds.), *Selective exposure to propaganda*. Chicago, IL: University of Illinois Press.

Shiffrin, R., & Schneider, W. (1977). Controlled and automatic human information processing: I. Detection, search and attention. *Psychological Review, 84* (1), 1-34.

Siemicki, M., Atkin, D., Greenberg, B., & Baldwin, T. (1987). Nationally distributed children's shows: What cable TV contributes. *Journalism Quarterly, 63* (4), 710-718.

Simon, E. (1985, January). Cable system profile: American cable of Phoenix. *Cable Marketing*, p. 45.

Singer, D., Zuckerman, D., & Singer, J. (1980). Helping young children learn about TV. *Journal of Communication, 30*, 84-93.

Sparkes, V. (1983a, November). *The people who don't subscribe to cable television: Who and why?* Paper presented at the annual meeting of MAPOR, Chicago.

Sparkes, V. (1983b). Public perception of and reaction to multi-channel cable television service. *Journal of Broadcasting, 27* (2), 163-175.

Sparkes, V. (1983c). The people who don't subscribe to cable television: Who and why? Paper presented at the annual meeting of MAPOR, Chicago, IL.

Sprafkin, J. N., & Silverman, L. T. (1981). Update: Physically intimate and sexual behavior on prime-time TV. *Journal of Communication, 31*, 34-40.

Stein, A. H., & Friedrich, L. A. (1972). Television content and young children's behavior. In J. Murray, E. Rubinstein, & G. Comstock (Eds.), *Television and social behavior (Vol. 2): Television and social learning*. Washington, DC: U.S. Government Printing Office.

Switchout yields 92 percent conversion of ex-spotlight subs. (1984, August). *Cable Marketing*, p. 12.

Television Audience Assessment, Inc. (1983). *The multi-channel environment*. Cambridge, MA: Author.

Thompson, J. W. (1986). *Flippers: Changes in the way Americans watch TV*. Chicago, IL: J. Walter Thompson Agency.

Twedt, D. W. (1964). How important to marketing strategy is the 'heavy user'? *Journal of Marketing, 28*, 71-72.

Villani, K. E. A. (1975). Personality/life style and television viewing behavior. *Journal of Marketing Research, 12*, 432-439.

Wakshlag, J., Agostino, D., Terry, H., Driscoll, P., & Ramsey, B. Television news viewing and network affiliation change. *Journal of Broadcasting, 27* (1), 53-68.

Wakshlag, J., & Greenberg, B. (1979). Programming strategies and the popularity of television programs for children. *Human Communication Research, 6* (1), 58-68.

Wakshlag, J., & Webster, J. (1986). *On explaining differences: Television viewing in broadcast, basic and pay cable households*. Unpublished manuscript.

Ward, S. (1974, September). Consumer socialization. *Journal of Consumer Research, 1*, 1-14.

Watt, J., & Krull, R. (1974). An information theory measure for television programming. *Communication Research, 1* (1), 44-68.

Webster, J. (1983). *The impact of cable and pay cable television on local station audiences.* Report for the National Association of Broadcasters, Washington, D.C.

Webster, J., & Agostino, D. (1982). *Cable and pay cable subscribers' viewing of public television stations.* Report of the Broadcast Research Center, Athens, Ohio.

Webster, J., & Wakshlag, J. (1983). A theory of program choice. *Communication Research, 10* (4), 430-447.

Wells, W. D. (1975). Psychographics: A critical review. *Journal of Marketing Research, 12,* 196-213.

Wells, W. D., & Tigert, D. J. (1971). Activities, interests, and opinions. *Journal of Advertising Research, 11* (4), 27-35.

Wentz, L. (1985, March 21). British VCR study shows most zap ads. *Electronic Media,* 18.

Zillmann, D. (1982). Television viewing and arousal. In D. Pearl, L. Bouthilet, & J. Lazar (Eds.), *Television and behavior: ten years of scientific progress and implications for the eighties (Vol. 2), Technical reviews.* Washington, DC: U.S. Government Printing Office.

Zillmann, D., & Bryant, J. (1985). Pornography, sexual callousness and the trivialization of rape. *Journal of Communication, 32* (4), 10-21.

Zillmann, D., Hezel, R. T., & Medoff, N. J. (1980). The effect of affective states on selective exposure to televised entertainment fare. *Journal of Applied Social Psychology, 10,* 323-339.

Zuckerman, M. (1979). *Sensation seeking: Beyond the optimal level of arousal.* Hillsdale, NJ: Erlbaum.

Author Index

Subject Index

A
Age, 17, 26, 70, 133–135, 215–216, 241–243

C
Cable subscribership, 207–225, 249–263, 264–288

Channel familiarity, 16, 21–22, 27, 197, 282–283

Channel loyalty, 36–37

Channel repertoire, 16–17, 24–25, 27, 37–39, 117–118, 146–147, 197, 217–223, 283–284, 296

Channel type, 56–59, 82–88, 91–96, 167–176

Children, 68–71, 89–96, 140–150, 153–163

Choice process models, 11, 25–29, 33–34, 47–48, 119–120, 131–136, 146–150, 197–198, 297–298

D
Daypart, 59–61, 171–173, 187–189

G
Group viewing, 42–43

Guide use, 21, 77, 97–98, 106–107, 146–147, 156–159, 217–223, 271–280

M
Movies, 162–165, 169–170, 172–176

MTV, 84, 95, 168, 237–245

N
New fall season, 74–88, 89–96

News, 162–165, 169–170, 171–173, 175, 179–190

Novelty-seeking, 17–19

O
Orienting search, 13–15, 22–23, 27, 39–41, 101–109, 117–118, 146–147, 155–156, 193–195, 217–223, 291–292

P
Pay cable, 17–18, 26, 43–44, 207–225, 226–236, 259, 270–288

Planning, 101–109, 117–118, 146–147, 153–154, 193–195, 217–223, 290–291

Playboy Channel, 191–203, 226–236

Program type, 35–36, 162–165

R
Radio listening style, 113–122

Reevaluation, 15–16, 23–24, 27, 41, 101–109, 117–118, 146–147, 159–162, 193–197, 217–223, 294–296

Remote control, 21, 44–47, 70, 215–216

S
Sex, 17–18, 26, 70, 151–166, 241–243

Sports, 162–165, 168, 169–170

V
Viewer availability, 35–39

Viewing style, 41–42, 53–54, 55–56, 71, 80–88, 91–96, 113–122, 129–139, 146–150, 292

Viewing time, 55–59, 82–85, 91–95, 292–294

Z
Zapping, 4, 62, 67–73, 195, see also reevaluation

Zipping, 4, 67